Quantum Chemical, Spectroscopic and Structural Study of Hydrochlorides, Hydrogens Squarates and Ester Amides of Squaric Acid of Amina

QUANTUM CHEMICAL, SPECTROSCOPIC AND STRUCTURAL STUDY OF HYDROCHLORIDES, HYDROGENS SQUARATES AND ESTER AMIDES OF SQUARIC ACID OF AMINA

TSONKO KOLEV

Nova Science Publishers, Inc.
New York

Copyright © 2008 by Nova Science Publishers, Inc.

All rights reserved. No part of this book may be reproduced, stored in a retrieval system or transmitted in any form or by any means: electronic, electrostatic, magnetic, tape, mechanical photocopying, recording or otherwise without the written permission of the Publisher.

For permission to use material from this book please contact us:
Telephone 631-231-7269; Fax 631-231-8175
Web Site: http://www.novapublishers.com

NOTICE TO THE READER

The Publisher has taken reasonable care in the preparation of this book, but makes no expressed or implied warranty of any kind and assumes no responsibility for any errors or omissions. No liability is assumed for incidental or consequential damages in connection with or arising out of information contained in this book. The Publisher shall not be liable for any special, consequential, or exemplary damages resulting, in whole or in part, from the readers' use of, or reliance upon, this material.

Independent verification should be sought for any data, advice or recommendations contained in this book. In addition, no responsibility is assumed by the publisher for any injury and/or damage to persons or property arising from any methods, products, instructions, ideas or otherwise contained in this publication.

This publication is designed to provide accurate and authoritative information with regard to the subject matter covered herein. It is sold with the clear understanding that the Publisher is not engaged in rendering legal or any other professional services. If legal or any other expert assistance is required, the services of a competent person should be sought. FROM A DECLARATION OF PARTICIPANTS JOINTLY ADOPTED BY A COMMITTEE OF THE AMERICAN BAR ASSOCIATION AND A COMMITTEE OF PUBLISHERS.

LIBRARY OF CONGRESS CATALOGING-IN-PUBLICATION DATA
Kolev, Tsonko.
  Quantum chemical, spectroscopic and structural study of hydrochlorides, hydrogens squarates and ester amides of squaric acid of amina / Tsonko Kolev.   p. cm.
  ISBN 978-1-60456-431-0 (hardcover)
  1. Organic chemistry. 2. Chemistry, Inorganic. 3. Amides. 4. Amines. I. Title.
  QD251.3.K65 2008
  547--dc22
                                  2008008017

Published by Nova Science Publishers, Inc. ✦ *New York*

CONTENTS

Preface		vii
Chapter 1	Introduction	1
Chapter 2	Results and Discussion	9
Chapter 3	Conclusion	79
References		83
Index		91

PREFACE

The interest in amino acid amides arises from their biologically important role. Some C-α-amidated amino acids Ile, Val, Thr, Ser, Met, Trp, Gln and Arg have been studied by single crystal X-ray diffraction and their bioactivity have been compared with the corresponding amino acids due to the fact that most of mammalian peptide hormones as calcitonin, gastrin, neurokinins or neuropeptides possess a C-α-terminal-amides. The C-α-amides are much more biologically active, comparing with the corresponding C-α-terminal free acids. For example the "potency ratio" of peptide amide to the corresponding peptide free acid in neurocinin is more than 40,000 to 1. Since the protonated forms of amino acid amides and C-α-amidated peptides exist in the living cell investigation could provide an understanding of their biological role. The choice of the acidity agent for the *in vitro* investigations is based manly on its own biological activity as for example squaric acid (H_2Sq). Its application for synthesis of optically active amino acid derivatives with potential nonlinear optical and electro-optical properties is well known, but its important biological role has been intensively studied in last five years. A large number of medications based on H_2Sq derivatives are effective inhibitors of protein tyrosine phosphatases or DNA polymerases from several viruses. H_2Sq diamides replaced a phosphate diester linkage in oligodeoxynucleotide. Selective antagonist of ionotropic glutamate receptors is obtained by replacing of γ-carboxylic acid of a glutamate residue within a polyamine toxin with squaric acid derivatives. Some H_2Sq-based peptides are inhibitors of matrix metalloprotease–1. These facts provoked the systematic investigations of hydrochlorides, hydrogensquarates and ester amides of squaric acid of amino acid amides of Ala, Arg, Tyr, Ser, Met, Ile, Lys, Tyr, Val, Leu, Pro and Phe. Some of them have been structurally characterized by single crystal X-ray diffraction. Their spectroscopic properties have been obtained using solid-state conventional and linear-polarized IR- and Raman spectroscopy and ^{1}H- and ^{13}C-NMR. However, the complicated spectroscopic data is difficult to a significant degree in their interpretation. Moreover, in the cases of hydrogensquarates and ester amides of squaric acid various intermolecular hydrogen bonds in solid-state with participation of H_2Sq have been established. Keeping in mind that physical and chemical properties of above mentioned compounds can be precisely calculated by means of ab initio and DFT methods at Hartee-Fock, MP2 and B3LYP level of theory, varying basis sets (6-31G*, 6-31G**, 6-31++G, 6-31++G*, 6-31++G**, 6-311G, 6-311G*, 6-311G** and 6-31++G**) have been employed. The results obtained allow a precise assignment of many vibrational bands to the corresponding normal modes. The electronic structure and conformational analysis have also been carried out.

Chapter 1

INTRODUCTION

The intermolecular hydrogen bonds are proposed to be the basis of the biological activities of molecules involved in biochemical processes in the living cell. These hydrogen bonds, with low binding energies of typically 10-30 kJ mol^1, allow the biomolecules to interact with their targets before breaking free [1]. Nearly all biochemically relevant molecules contain intramolecular hydrogen bonds. Therefore, the *in vivo* biochemical processes involve a combination of intra- and intermolecular hydrogen bonding interactions, the details of which are not always well understood.

Amino acids are the simplest biomolecules that contain intramolecular hydrogen bonds, and they are building blocks of the peptides and proteins. Structure determination and the identification of intramolecular hydrogen bonding motifs in these systems may provide insight into the interactions in larger systems as well as to provide valuable experimental data for testing and refining quantum chemical methods. However, polypeptides in the living cell are joined together by amide linkages and the intramolecular hydrogen bonding networks in peptides are better represented by amino acid amides than amino acids, and the amino amide derivatives may be described as simple peptide models. Moreover, it is known that most of mammalian peptide hormones possess a C-α-terminal-amide as exemplified by calcitonin, gastrin, neurokinins, neuropeptides and their related peptides [2]. In many cases C-α-amides are much more biologically active in comparison with the corresponding C-α-terminal free acids. The importance of the C-α-terminal amide for bioactivity can be illustrate by the "potency ratio" which is defined as the relative potency of a peptide amide divided by the relative potency of the corresponding peptide free acid e.g. neurocinin A possesses potency ratio > 40 000 [3,4]. At present the biological-structural function of the

C-α-amide is not fully understood. The amidation arises from the oxidative cleavage of C-α-terminal glycine-extended prohormones [5]. Since the protonated form of amino acid amide exist in the living cell, such information may be useful on understanding the different biological function between C-α-amide and C-α-acid peptides. For these reasons, a determined the conformational structures of alaninamide [6], prolinamide [7] and valinamide [1] has been presented. In contrast to the simple amino acids, only one gas phase conformation has been observed for each of these molecules. The backbone structure is stabilized by an intramolecular hydrogen bond from an amide proton to the amino nitrogen, and this conformation is unchanged in the alaninamide-water van der Waals complex. The structural study on some C-α-amidated amino acids Ile, Val, Thr, Ser, Met, Trp, Gln, Arg [8] and Tyr [9] have been performed and compared with their C-α-unamidated counterparts [8].

On the other hand the bioactive role not only of the neutral peptides as well as of their protonated forms is also known connected with the different possible conformations of peptide chains. Moreover, most of the *in vivo* conditions are characterized with acidity of the medium, resulted to protonated active form of peptides and proteins. The choice of the acidity agent for the *in vitro* investigations arise from its own bioactive function. Squaric acid is an alternative for studying of the protonated forms of amino acid amides and peptides. Its application for synthesis of optically active derivatives of amino acids, with potential nonlinear optical and electro-optical properties is known [10-14]. However, its important biorole is intensively studied in last decade. Large number of medications based on squaric acid derivatives are obtained [15,16]. Some amides are effective monoanionic inhibitors of protein tyrosine phosphatases [17], which are important targets in medicinal chemistry. These enzymes play role in the number of human diseases as diabetes type II an infection by *Yersinia pestis* [17]. Other, are selective inhibitors of DNA polymerases from several viruses [18]. The squaric acid amides of antracycline glycoside-type antibiotics as daunomycin, adriamycin, epirubicin and carminomycin are potential antitumor agents as far as the corresponding nonsubstituted drugs are widely used in cancer chemotherapy [18]. The usage of squaric acid diamides to replace a phosphate diester linkage in an oligodeoxynucleotide is also reported [15,18]. A selective antagonist of ionotropic glutamate receptors is obtained by replacing of γ-carboxylic acid of a glutamate residue within a polyamine toxin with squaric acid derivatives [15,16,17]. Other products are applied to investigation of an NMDA antagonist that regulated the activation of glutamate receptors. A novel series of benzylamine derivatives of squaric acid are potential potassium channel operators for treatment of urge urinary incontinence [17]. The replacement of the amino carboxylic moiety in thioproline CT5219, which is a Very Late Antigen-4 antagonist with squaric acid, is potent medication for treatment of asthma, multiple sclerosis and rheumatoid arthritis [19]. Some squaric acid-based peptide conjugates are evaluated as inhibitors of matrix metalloproteases [20], which regulate structure and sustain a balanced composition of extra cellular matrix. Last processes are important for maintaining normal physiology in a number of tissues [19]. Some derivatives of β-lactam antibiotics such as penicillin and cephalosporins are with potencial usage for interfering with the biosynthesis of the peptidoglycan layer of bacterial cell like non-substituted antibiotics [21]. The neurocemical activity of some derivatives is also reported [22]. In the same time a series of amino-acid derivatives of squaric acid are also demonstrated [9,10,11,23-31] due to the protonated forms of amino acid amides represent a new class of compounds, having great biological importance. The systematic investigations of amino acid amides and their derivatives with squaric acid and diethylsqaurate is required for understanding of the properties of di-, tri-, tetra- and polypeptides, which are the fundamental role in the bioactive properties *in vivo*. The mentioned model systems described adequately the peptide properties. During the last two years series of structural studies on amino acid amides, including hydrogensquarates of prolinamide [25], tyrosinamide [9], argininamide [31], leucinamide and ester amides of squaric acid of phenylalaninamide [23], methioninamide [27,32], alaninamide [24], prolinamide [25], valinamide [28] and tryptophanamide [26] have been carried out. In all these cases the role of intra and intermolecular interactions on the vibrational (Raman and IR-) spectra in solid phase has been estimated as well as some stereo structural predictions have been made using the possibilities of linear-polarized IR-spectroscopic approach of oriented solid-samples as suspensions in nematic liquid crystal (see below). The conformational preferable in gas phase for the protonated tyrosinamide [33] and ester amides of squaric acid of alaninamide [34], methioninamide [32] and valinamide [28] have been studied by ab initio and DFT calculations. The neutral forms of valinamide, prolinamide and alaninamide have been examining theoretically by UMP2/6-31G** in [1,6,7] and the experimental single crystal X-ray of $_L$-leucinamide in [35]. In last research a complex structural determination has been supported as well with solid-state IR- and NMR data [35] and has been established two magnetically non-equivalent molecules in the unit cell of crystalline $_L$-leucinamide. This phenomenon has been proposed in the non-polarized FTIR spectra as well [35]. With the possibilities of solid-state linear-polarized IR-spectroscopy this phenomenon in the optical IR-spectra is experimentally observed and confirmed (see below).

Hence, the paper contains the conformational and structural investigation of hydrochlorides, hydrogensquarates and ester amides of squaric acid of amides of the following amino acids: alanine, arginine, isoleucine, leucinee, methionine, phenylalanine, threonine, tryptophan, tyrosine, prolin, serine and valine (Scheme 1).

Argininamide dihydrochloride (*ArgNH$_2$.2HCl*).

Isoleucinamide hydrochloride (*IleNH$_2$.HCl*).

Methioninamide hydrochloride (*MetNH$_2$.HCl*).

Serinamide hydrochloride (*SerNH₂.HCl*).

Tyrosinamide hydrochloride (*TyrNH₂.HCl*).

Tryptophanamide hydrochloride (*TrpNH₂.HCl*).

Threoninamide hydrochloride (*ThrNH₂.HCl*).

Valinamide hydrochloride (*ValNH₂.HCl*).

Alaninamide hydrogensquarate (*AlaNH₂.HSq*).

Argininamide bis(hydrogensquarate) (*ArgNH₂.2HSq*).

Prolinamide hydrogensquarate (*ProNH₂.HSq*).

Tyrosinamide hydrogensquarate (*TyrNH₂.HSq*).

Methioninamide ester amide of squaric acid (*MetNHSqEt*).

Phenylalanin amide ester amide of squaric acid (*PheNHSqEt*).

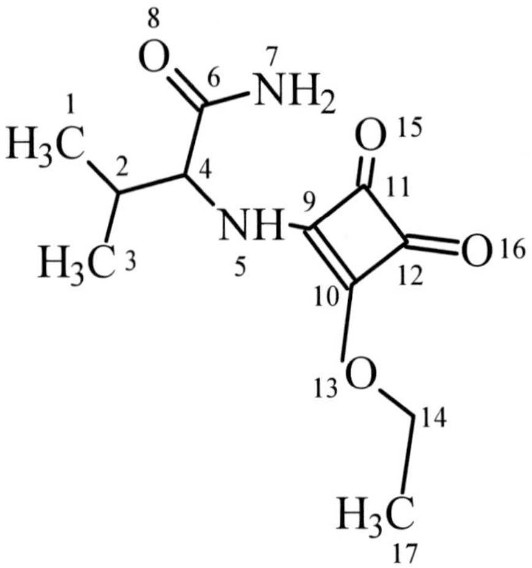

Valinamide ester amide of squaric acid (*ValNHSqEt*).

Scheme 1. Chemical formula and abbreviations of the studied amino acid amide derivatives.

The theoretical results are compared with experimental ones obtained by single crystal X-ray diffraction with a view to explain the role of intra- and intermolecular hydrogen bonds on the conformational preference in solid-state comparing with gas phase data. In the same aspect are described the solid-state IR-LD spectra of oriented samples as suspension in nematic liquid crystal (see below) of compound studied, assigning the vibrational bands to the corresponding normal modes.

Chapter 2

RESULTS AND DISCUSSION

HYDROCHLORIDES AND HYDROGENSQUARATES

The quantum chemical calculations were performed with Dalton 2.0 [36] and GAUSSIAN-98 program packages [37]. The output fails are visualized by means of ChemCraft program [38]. The geometries of all the compounds (scheme 1) are optimized at level of theory: Restricted and Unrestricted Hartree-Fock (RHF and UHF), second-order Moller-Pleset perturbation theory (UMP2 and MP2), using basis sets as 6-31G*, 6-31G**, 6-31++G, 6-31++G*, 6-31++G**, 6-311G, 6-311G*, 6-311G** and 6-311++G** [39-62]. The density function theory (DFT) method employed is B3LYP, which combines Becke's three-parameter nonlocal exchange with the correlation function of Lee, Yang and Parr [63,64] is also applied. Molecular geometries of the studied species were fully optimized by the force gradient method using Bernys' algorithm [65]. The conformational analysis in gas phase is carried out by follow way. To generate the (φ,χ) potential energy surface, the structures of protonated compounds are calculated at the *ab initio* UHF/6–31G*//UHF/6-311++G** or RHF/6–31G*//RHF/6-311++G** levels. In each structure, all geometrical parameters were fully relaxed, except for the constrained torsion angles φ and χ_1. Values of these angles were chosen by using a step size of 0.5°, within the range from $-180°$ to $180°$ [66]. The minima observed on the hypersurface were then subjected to full geometry optimization at the DFT/B3LYP/6-311++G** level or UHF/6-311++G**, RHF/B3LYP/6-311++G**, UMP2/6-311++G** and MP2/6-311++G**, which should enable correct prediction of the stability order of the minima calculated [67], followed by a second derivative analysis (frequency), which proved all of them to be minima. The absence of the imaginary frequencies, as well as of negative eigenvalues of the second-derivative matrix, confirmed that the stationary points correspond to minima of the potential energy hypersurfaces. The geometry parameters of the corresponding energy-minimized conformers were then further discussed. The accessible conformational space of the molecule studied was assumed on the basis of the close resemblance between the Ramachandran contact map and the energy contours map within the limit of 5.0 kcal.mol^{-1} [68], as is also applied elsewhere [69,70]. In some cases the limit of 13 kcal.mol^{-1} is used as well. The space was calculated, using the radial basis function as a girding method. As the overall conformational profiles of modified systems can differ from those of common peptide models, we describe the energy-minimized conformers of the investigated molecule by the general short hand letter notation introduced by Zimmerman [69]. The torsion angle φ is determined by the relatively disposition of cationic NH_3^+ and electronegative carbonyl atom. The χ_1 is definite as $N(N^+H_3)_i$-C_j-C_k-C_l, connected with the conformations in amino acid amide side chains. In tables 2-11 are summarized the geometry parameters corresponding to most stable conformers of given protonated form.

The calculations of protonated form of argininamide are obtained according the following statement (Scheme 2). Opposite to other amino acids, stabilizing an H_3N^+-R-COO$^-$ form, in the arginine-containing systems, the significant basic properties of guanidyl fragment leads to an intramolecular hydrogen-transfer and formation of (**2**) (Scheme 2). A next charge redistribution and an equalization of bond lengths in guanidyl-fragment are established (Scheme 2 (**3**)). Then the protonation leads to obtaining of NH_3^+ fragment in the structure. In the case of corresponding amide derivative the protonation of form (**4**) resulted to obtaining of NH_3^+ group shown as (**5**) in the same Scheme 2.

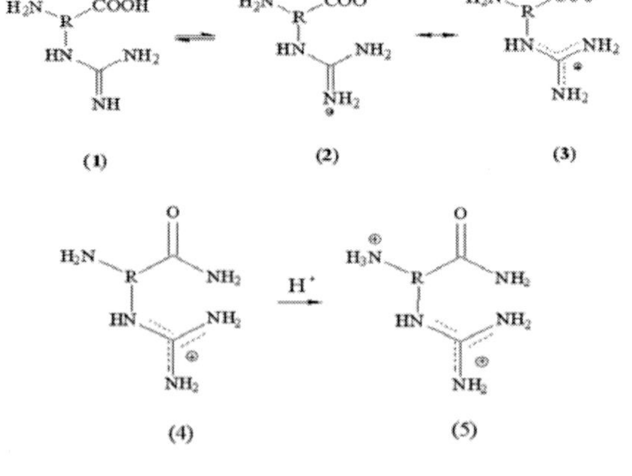

Scheme 2. Resonance structures and protonated forms of arginine and argininamide.

The calculations identified a series of conformational minima for each of the protonated form studied and the following numbers of which lie lower than 13.0 kJ/mol: 4 (*ArgNH$_2$*), 2 (*IleNH$_2$*), 7 (*MetNH$_2$*), 1 (*SerNH$_2$*), 3 (*TyrNH$_2$*), 2 (*TrpNH$_2$*), 4 (*ThrNH$_2$*), 3 (*ValNH$_2$*), 8 (*AlaNH$_2$*) and 4 (*ProNH$_2$*), respectively. Backbone conformations with intramolecular $N^+H_3\ldots O=C-NH_2$ hydrogen bonds were found to be more stable than structures with bifurcated amine (N^+H_3) to carbonyl (C=O) interactions. The former bond is a conventional interaction commonly accepted to play the dominant role in determining the relative stability of the conformation. Table 1 list the relative energies, not corrected for zero-point energies, for the most stable conformers of each of the compound studied. Like in the case of neutral valinamide [1], three orientations of the isopropyl group (with φ and χ_1 approximately 33.0°, 55.0° (E_{rel} = 9.0 kcal/mol); 27.0°, 172.5° (E_{rel} = 6.2 kcal/mol) and 40.0°, 285.5o (E_{rel} = 11.3 kcal/mol)) were found (Scheme 3) in protonated *ValNH$_2$*. In the case of protonated *ProNH$_2$* containing five-membered ring restricts torsions about the bond between C_3 (scheme 1) and the amino nitrogen. Non-substituted proline reduces the flexibility of peptides and is often found in the bends and kinks of folded protein chains. Despite proline's unique connectivity, HF/4-21G [71] and HF/6-31G [72] calculations find that the most stable conformation contains an intramolecular hydrogen bond from the amine to the carbonyl oxygen and is similar to the lowest energy glycine and alanine conformers [7]. The conformations of prolin and prolin derivatives are complicated by puckering of the pyrrolidine ring, which relieves torsion strain caused by eclipsing methylene groups on the ring edges. For these seasons, like to neutral prolinamide the conformational analysis of protonated the conformations were also modeled at the Hartree-Fock level (6-31G*) (figure 1 and see also table 11) and the optimized structure converged to a C_2-endo (Scheme 4) envelope ring conformation with φ = 0°. Two twist structures and an additional envelope conformation was found within 9 kcal mol^{-1} of the minimum energy structure. These structures may not represent local minima on the potential energy surface since φ and the ring torsion angles were constrained as described above. The model calculations reveal that there are many different low-energy conformations, although the experimental spectrum may not correspond to the lowest energy structure.

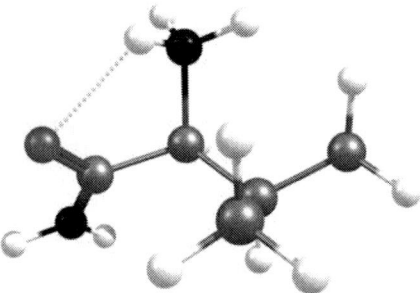

(E_{rel} = 5.0 kcal/mol).

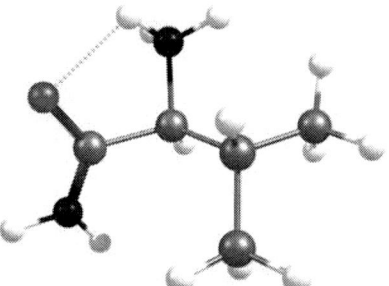

(E_{rel} = 2.2 kcal/mol).

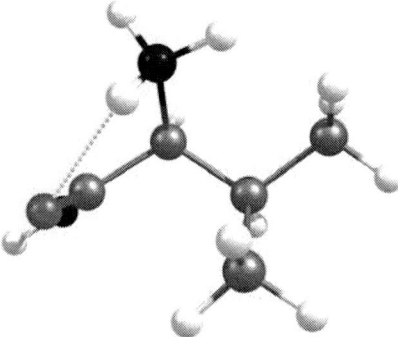

(E_{rel} = 4.3 kcal/mol).

Scheme 3. Three conformers (C_i, i = 1-3) of protonated *ValNH$_2$* after optimization at UHF/6-311++G** level of theory.

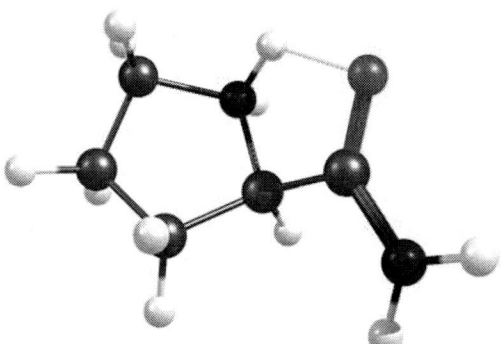

C_1 (E_{rel} = 1.7 kcal/mol).

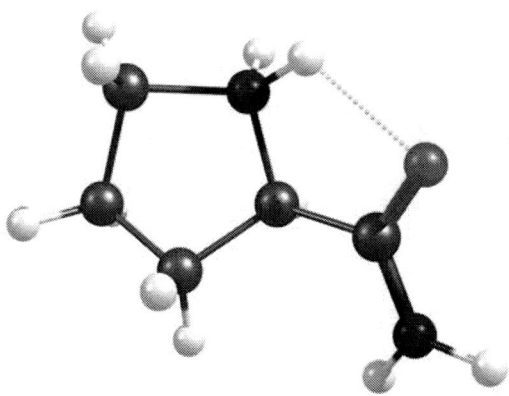

C_2 (E_{rel} = 8.7 kcal/mol).

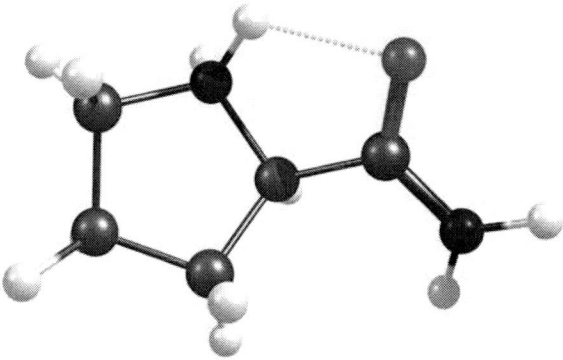

C_3 (E_{rel} = 0.6 kcal/mol).

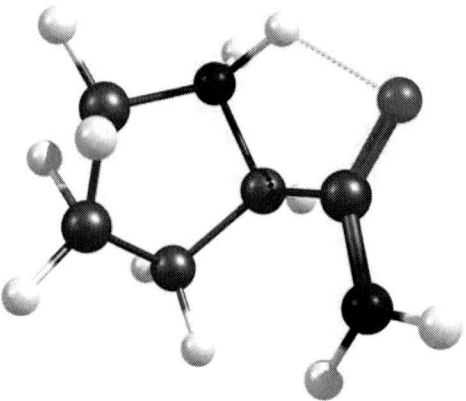

C_4 (E_{rel} = 9.0 kcal/mol).

Scheme 4. Ab initio model conformers (C_i, i = 1-4) of protonated *ProNH₂* after optimization at UHF/6-31G* level of theory.

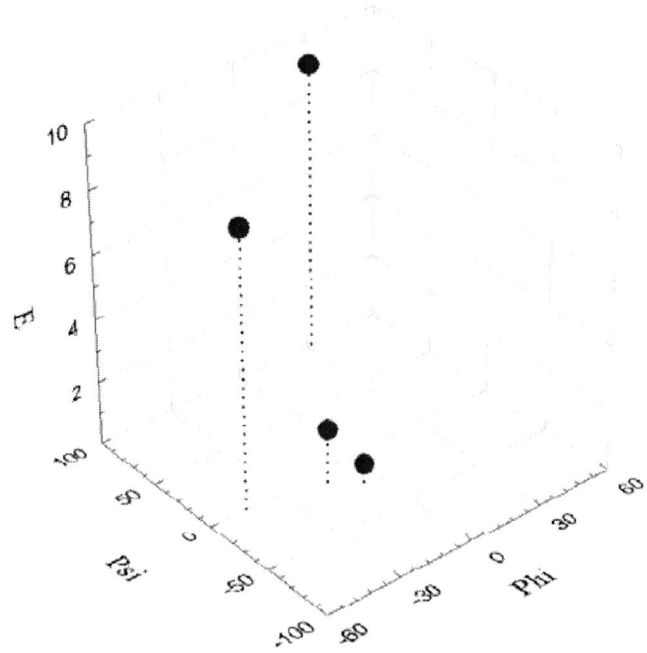

Figure 1. Ab initio model conformers of protonated *ProNH₂* at UHF/6-31G*.

The predicted $N^+H_3\ldots O=C-NH_2$ bond lengths of protonated amino acid amides varying within 2.550 – 2.528 Å (Schemes 5-11) and the corresponding $(H_3)N^+-H-O(=C)$ angles of 114.7(4)° (*ArgNH$_2$*), 117.4(4)° (*IleNH$_2$*), 115.1(6)° (*MetNH$_2$*), 119.2(3)° (*SerNH$_2$*), 112.3(6)° (*TyrNH$_2$*), 117.2(8)° (*TrpNH$_2$*), 118.7(6)° (*ThrNH$_2$*), 112.6(9)° (*ValNH$_2$*), 115.3(4)° (*AlaNH$_2$*) and 118.1(9)° (*ProNH$_2$*) are obtained, respectively.

Table 1. Ab initio most stable (UHF/6-311++G**) conformers of protonated amino acid amides

Compound	φ	χ$_1$	Rel. energy [kcal/mol]
ArgNH$_2$	23.5	74.5	0.0
IleNH$_2$	15.0	73.5	2.8
MetNH$_2$	5.05	47.0	4.9
SerNH$_2$	8.0	173.5	7.0
TyrNH$_2$	17.5	54.0	3.3
TrpNH$_2$	14.0	174.0	4.5
ThrNH$_2$	8.0	130.0	3.5
ValNH$_2$	27.7	172.5	2.2
AlaNH$_2$	15.0	-	3.5
ProNH$_2$	6.0	24.6	0.6

In all theoretical approximated molecules, the amide $NH_2-C=O$ fragment is flat with maximal deviation of the planarity within 0.5° – 1.2° as well as the guanydyl-fragment in protonated argininamide is planar with torsion angles deviation within 2.0(9)° - 3.6(5)° range (Scheme 5). The assumption about the partial singly charge redistribution in last fragment is supported by the theoretical obtained values of +1.56,+ 1.43 and + 0.8 eV for both primary and the one secondary amino groups in gyanidyl fragment by means of generalized atomic polar tensor (GAPT) model (see ref. 73). The similar result has been obtained for arginine containing di- and tetrapeptides [74].

The role on the used level of theory and basis set on the predicted geometry parameters in these class of organic compounds have been evaluated on the examples of protonated alaninamide [30], threoninamide [8], leucinamide [75] and prolinamide [29] (tables 2 - 11). As may be see, independently of the used level of theories and basis sets, in all cases a insignificant differences of the experimental data are established towards bond lengths and angles (tables 2-11) determining that the data do not differ by more than 0.093 Å and 7.8(9)° [8,29,30,75], which is a good approximation, illustrating the applicability of these theoretical approaches for structural prediction in gas phase. However, the theoretical predicted dihedral angles in all cases varies with deviation towards the experimental single crystal X-ray diffraction data within 2 – 25% (see Schemes 2 - 10 and tables 2 - 11). This result is explained with the preferable intermolecular interactions in solid phase between separate amino acid amides in the frame of the unit cell or between them, the anionic species and included solvent molecules. The single crystal X-data diffraction results of protonated amino acid amides indicate the preferable intermolecular interactions, and an absence of any intramolecular ones.

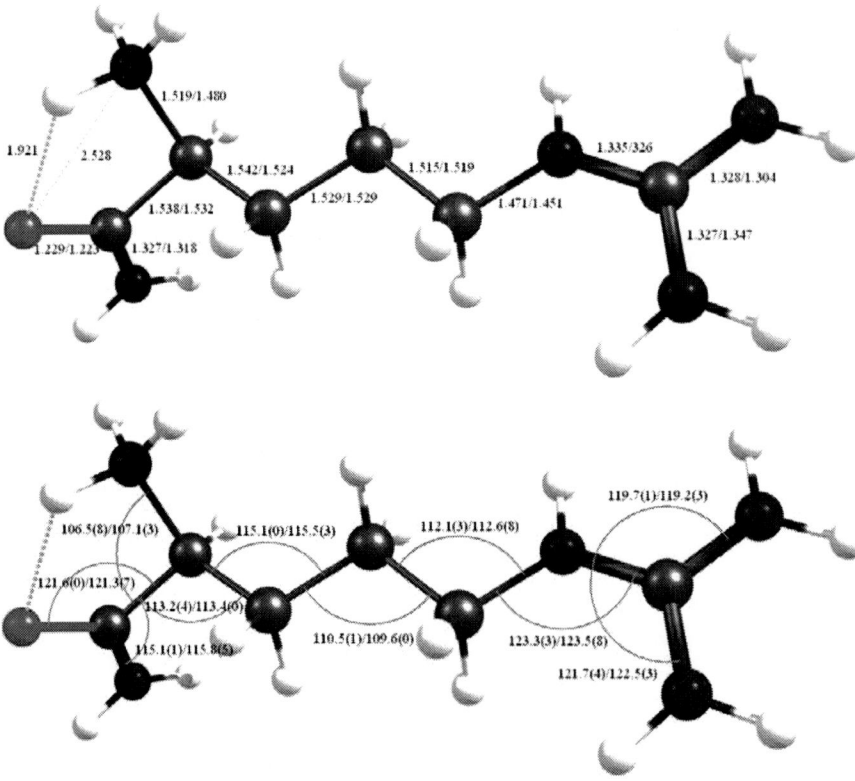

Scheme 5. Ab initio (UHF/6-311++G**)/experimental (ref. 31) data of bond lengths [Å] and angles [°] of protonated *ArgNH₂*.

Results and Discussion 17

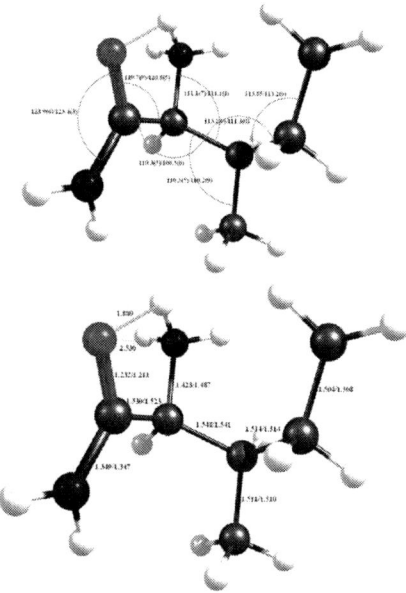

Scheme 6. Ab initio (UHF/6-311++G**)/experimental (ref.8) data of bond lengths [Å] and angles [°] of protonated *IleNH₂*.

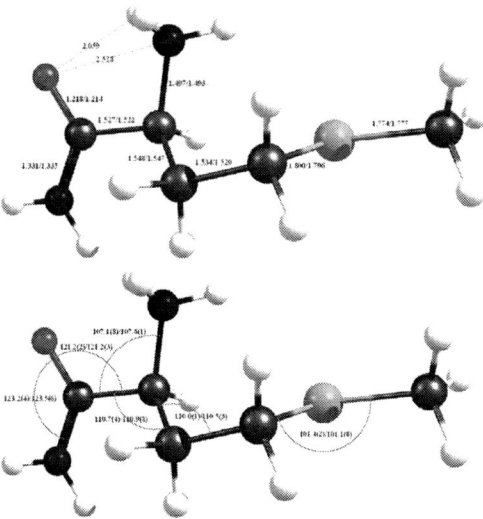

Scheme 7. Ab initio (UHF/6-311++G**)/experimental (ref.8) data of bond lengths [Å] and angles [°] of protonated *MetNH₂*.

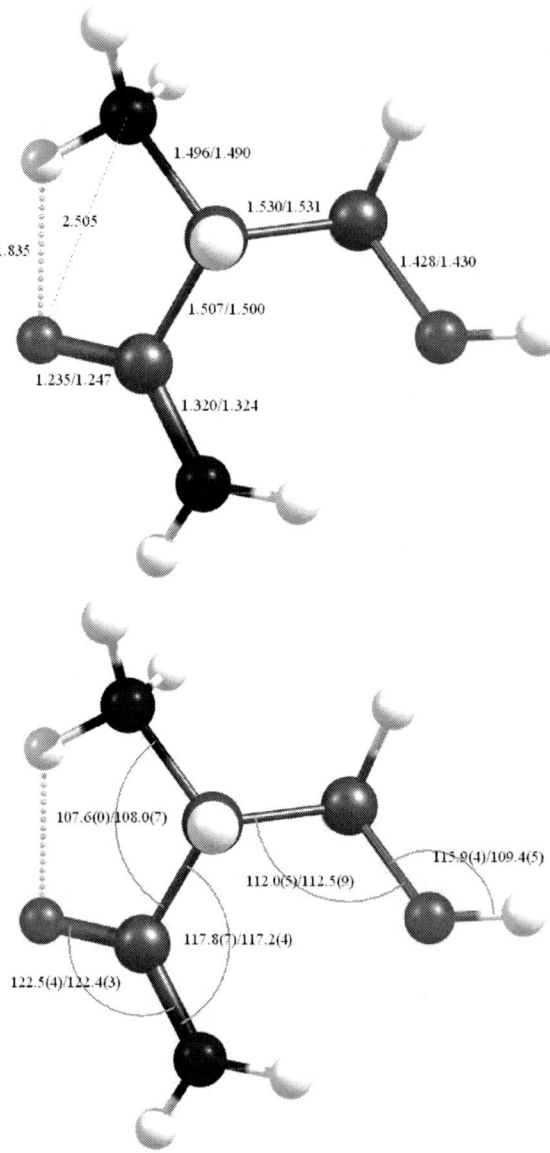

Scheme 8. Ab initio (UHF/6-31++G**)/experimental (ref.8) data of bond lengths [Å] and angles [°] of protonated *SerNH₂*.

Results and Discussion 19

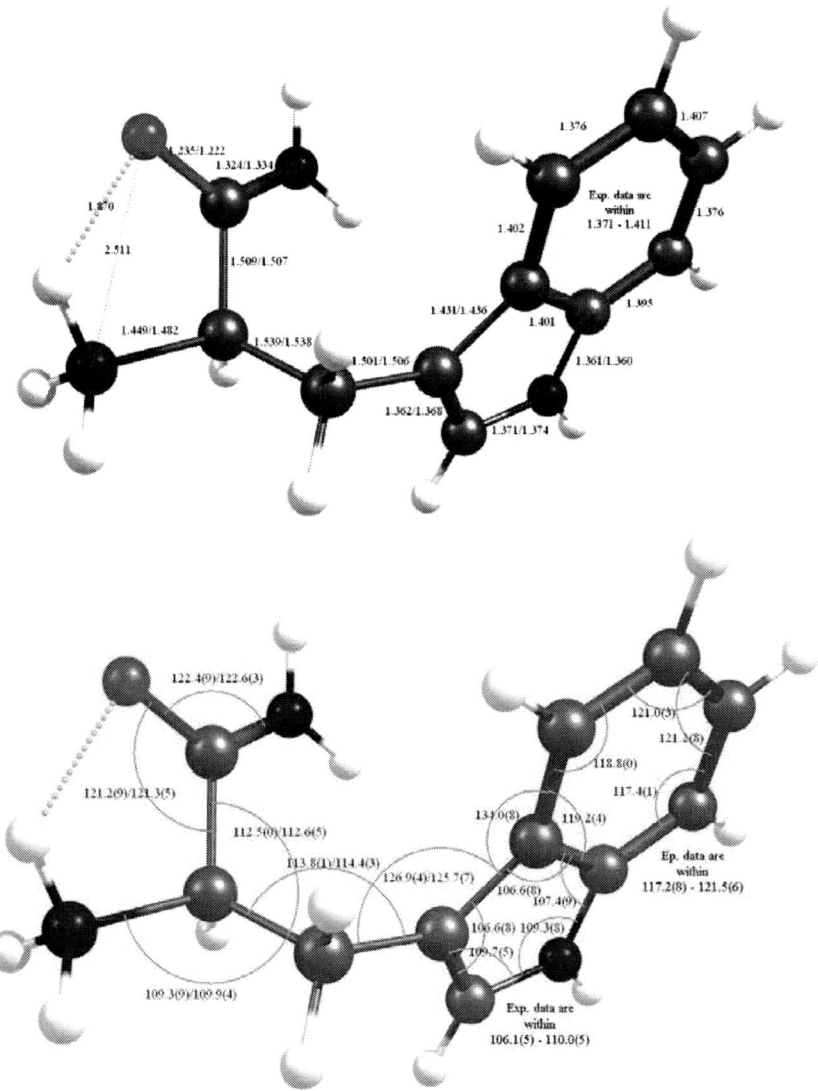

Scheme 9. Ab initio (UHF/6-311++G**)/experimental (ref.8) data of bond lengths [Å] and angles [°] of protonated *TrpNH₂*.

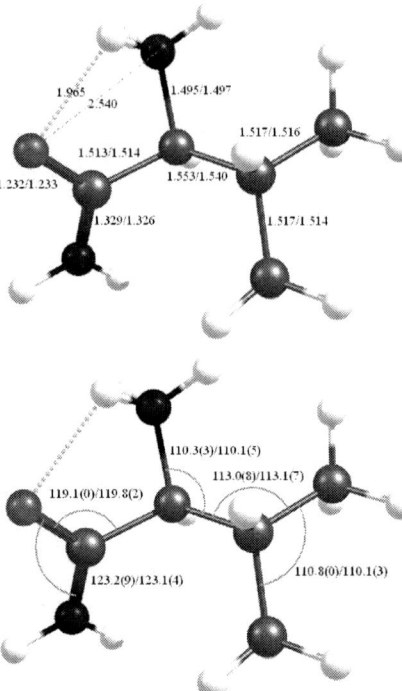

Scheme 10. Ab initio (UHF/6-311++G**)/experimental (ref.8) data of bond lengths [Å] and angles [°] of protonated *ValNH₂*.

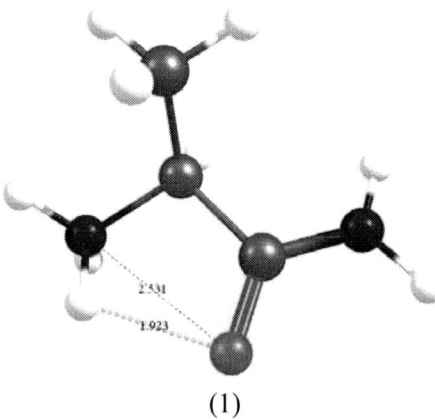

(1)

Scheme 11. (Continued on next page.)

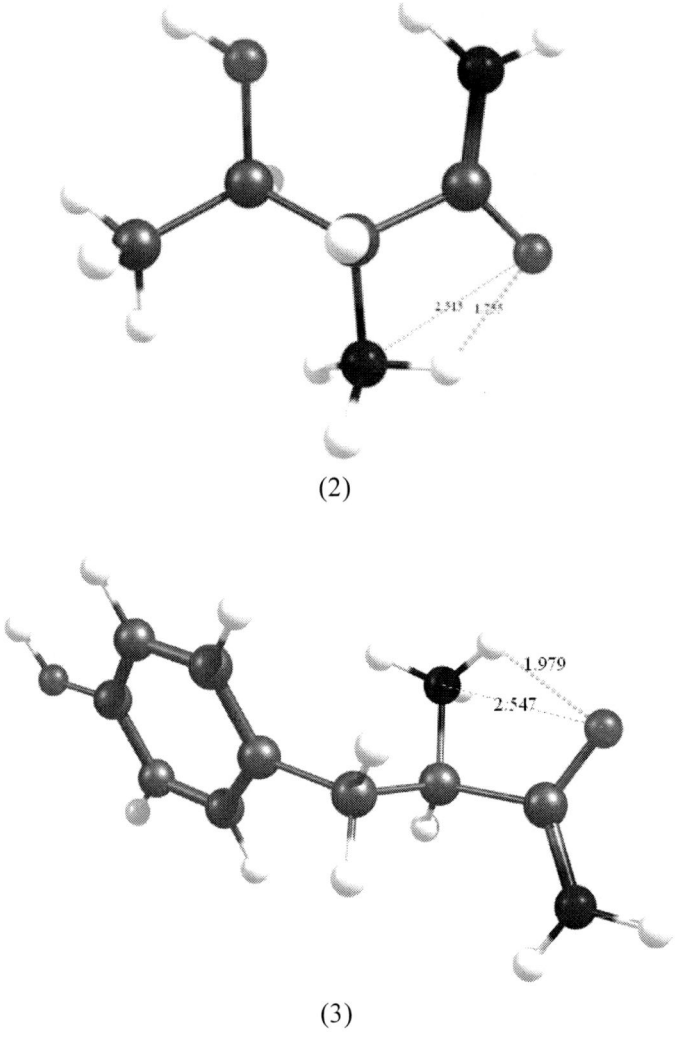

Scheme 11. Predicted (UHF/6-311++G**) intramolecular hydrogen bonding in protonated *AlaNH$_2$* (1), *ThrNH$_2$* (2) and *TyrNH$_2$* (3).

Table 2. Calculated and experimental dihedral angles of protonated ArgNH$_2$, using atom numbering Scheme 1

Name definition	Dihedral angles [°]		
	UHF/6-311++G**	Experimental data	
		Ref. [8]	Ref. [31]
D(1,2,4,5)	178.7	178.6	178.9
D(3,2,4,5)	1.0	0.0	0.8
D(2,4,5,6)	174.8	89.3	132.8
D(4,5,6,7)	178.0	178.5	58.3
D(5,6,7,8)	177.4	174.7	175.5
D(6,7,8,9)	171.4	55.8	161.8
D(6,7,8,10)	74.2	66.34	79.3
D(7,8,9,11)	95.5	119.6	97.8
D(7,8,9,12)	82.5	59.8	79.2
D(10,8,9,11)	23.5	5.3	23.9
D(10,8,9,12)	158.3	175.2	157.9

Table 3. Calculated and experimental dihedral angles of protonated IleNH2 using atom numbering Scheme 1

Name definition	Dihedral angles [°]	
	UHF/6-311++G**	Ref. [8]
D(1,2,3,4)	169.6	166.0
D(1,2,3,5)	64.6	70.6
D(2,3,5,6)	73.7	62.6
D(2,3,5,7)	44.3	178.9
D(4,3,5,6)	160.4	173.2
D(4,3,5,7)	81.3	78.7
D(3,5,7,8)	74.3	110.2
D(3,5,7,9)	106.0	68.2
D(6,5,7,8)	163.6	129.5
D(6,5,7,9)	15.9	51.9

Table 4. Calculated and experimental dihedral angles of protonated MetNH2, using atom numbering Scheme 1

Name definition	Dihedral angles [°]	
	UHF/6-311++G**	Ref. [8]
D(1,2,3,4)	173.9	77.9
D(2,3,4,5)	43.9	179.6
D(3,4,5,6)	166.7	71.4
D(3,4,5,7)	47.3	170.3
D(4,5,6,8)	53.5	99.5
D(4,5,6,9)	127.8	84.8
D(7,5,6,8)	176.2	147.8
D(7,5,6,9)	5.0	33.8

Table 5. Calculated and experimental dihedral angles of protonated SerNH2 using atom numbering in Scheme 1

Name definition	Dihedral angles [°]	
	UHF/6-311++G**	Ref. [8]
D(1,2,3,4)	173.0	75.7
D(1,2,3,5)	67.7	44.5
D(2,3,5,6)	131.5	101.5
D(2,3,5,7)	50.0	76.6
D(4,3,5,6)	8.8	18.6
D(4,3,5,7)	172.7	161.6

Table 6. Calculated and experimental dihedral angles of protonated TrpNH2 using atom numbering Scheme 1

Name definition	Dihedral angles [°]	
	UHF/6-311++G**	Ref. [8]
D(6,1,2,3)	0.1	0.4
D(2,1,6,5)	0.2	0.9
D(2,1,6,7)	179.4	179.4
D(1,2,3,4)	0.2	0.6
D(2,3,4,5)	0.0	1.0
D(3,4,5,6)	0.2	0.5
D(3,4,5,11)	179.3	179.5
D(4,5,6,1)	0.4	0.4
D(4,5,6,7)	179.7	179.8
D(11,5,6,1)	179.3	179.6
D(11,5,6,7)	0.1	0.2
D(6,5,11,8)	0.5	0.1
D(1,6,7,8)	178.6	179.4
D(1,6,7,9)	1.7	0.7
D(5,6,7,8)	0.6	0.4
D(5,6,7,9)	177.4	179.5
D(6,7,8,11)	0.9	0.5
D(9,7,8,11)	177.8	179.4
D(6,7,9,10)	100.6	102.3
D(8,7,9,10)	83.1	77.7
D(7,8,11,5)	0.9	0.3
D(7,9,10,12)	74.0	57.7
D(7,9,10,13)	69.9	62.8
D(9,10,13,14)	75.6	64.8
D(9,10,13,15)	104.9	111.8
D(12,10,13,14)	165.1	173.6
D(12,10,13,15)	14.2	9.7

Table 7. Calculated and experimental geometry parameters of protonated TyrNH2 using atom numbering Scheme 1

Name definition	Bond lengths [Å] and angles [°]		
	MP2/6-311++G**	Experimental data	
		Ref. [9]	Ref. [76]
R(1,2)	1.378	1.391	1.381
R(1,7)	1.399	1.390	1.391
R(2,3)	1.393	1.378	1.394
R(3,4)	1.361	1.378	1.356
R(3,5)	1.384	1.384	1.391
R(5,6)	1.390	1.395	1.380
R(6,7)	1.391	1.392	1.393
R(7,8)	1.500	1.517	1.495

Table 7. (Continued)

Name definition	Bond lengths [Å] and angles [°]		
	MP2/6-311++G**	Experimental data	
		Ref. [9]	Ref. [76]
R(8,9)	1.549	1.539	1.545
R(9,10)	1.526	1.528	1.526
R(9,11)	1.487	1.492	1.483
R(10,12)	1.330	1.313	1.313
R(10,13)	1.231	1.222	1.233
A(2,1,7)	121.0(6)	121.2(3)	121.9(9)
A(1,2,3)	119.0(0)	119.4(1)	118.9(0)
A(2,3,4)	116.2(9)	117.7(3)	118.9(7)
A(2,3,5)	120.2(1)	120.7(7)	120.2(4)
A(3,5,6)	119.5(4)	119.2(4)	119.5(0)
A(1,7,8)	120.9(3)	121.6(6)	120.4(5)
A(7,8,9)	112.7(0)	112.9(6)	115.1(4)
A(8,9,10)	111.8(6)	111.1(1)	112.1(6)
A(8,9,11)	109.5(8)	109.0(0)	111.5(0)
A(9,10,12)	117.6(5)	116.4(9)	116.0(2)
A(12,10,13)	124.5(5)	124.0(0)	124.7(6)
D(7,1,2,3)	0.2	0.5	1.3
D(2,1,7,6)	0.6	0.4	1.2
D(2,1,7,8)	179.9	179.8	179.6
D(1,2,3,4)	179.7	179.6	178.7
D(1,2,3,5)	0.1	0.5	0.9
D(1,7,8,9)	70.8	83.4	83.9
D(7,8,9,10)	170.2	72.8	52.7
D(7,8,9,11)	54.1	170.1	67.9
D(8,9,10,12)	76.7	105.6	76.5
D(8,9,10,13)	101.6	73.4	101.2
D(11,9,10,12)	164.5	135.8	160.8

Table 8. Calculated and experimental geometry parameters of protonated ThrNH2 using atom numbering Scheme 1

Name definition	Bond lengths [Å] and angles [°]			
	MP2/6311++G**	UHF/6311++G**	UHF/631++G**	Ref. [8]
R(1,2)	1.511	1.512	1.523	1.511
R(2,3)	1.419	1.417	1.421	1.415
R(2,4)	1.523	1.525	1.527	1.523
R(4,5)	1.532	1.532	1.535	1.531
R(4,6)	1.488	1.489	1.492	1.487
R(5,7)	1.322	1.324	1.327	1.321
A(1,2,3)	108.0(3)	108.0(8)	110.9(2)	108.2(8)
A(1,2,4)	111.2(9)	111.3(6)	112.8(0)	111.0(5)
A(2,4,5)	112.3(5)	112.3(6)	114.7(3)	112.5(9)

Table 8. (Continued).

Name definition	Bond lengths [Å] and angles [°]			
	MP2/6311++G**	UHF/6311++G**	UHF/631++G**	Ref. [8]
A(2,4,6)	110.6(3)	110.7(9)	111.6(7)	110.6(0)
A(4,5,7)	115.3(3)	115.7(2)	116.7(1)	115.3(8)
D(1,2,4,5)	175.2	174.8	174.3	172.9
D(1,2,4,6)	55.9	55.3	56.6	66.2
D(3,2,4,5)	65.2	65.3	66.3	67.3
D(3,2,4,6)	175.3	175.0	176.0	53.5
D(2,4,5,7)	49.3	49.3	51.3	75.8
D(2,4,5,8)	132.7	132.5	130.1	103.3
D(6,4,5,7)	172.6	172.9	173.6	161.7
D(6,4,5,8)	9.4	8.9	7.8	19.0

Table 9. Calculated and experimental dihedral angles of protonated ValNH2 using atom numbering Scheme 1

Name definition	Dihedral angles [°]	
	UHF/6-311++G**	Ref. [8]
D(1,2,4,5)	172.4	171.3
D(1,2,4,6)	57.9	68.1
D(3,2,4,5)	64.8	63.5
D(3,2,4,6)	57.4	57.5
D(2,4,6,7)	90.1	80.7
D(2,4,6,8)	90.5	97.3
D(5,4,6,7)	151.5	157.0
D(5,4,6,8)	27.7	25.1

Table 10. Calculated and experimental geometry parameters of protonated AlaNH2 using atom numbering in Scheme 1

Name definition	Bond lengths [Å] and angles [°]		
	UHF/6-311++G**	UHF/6-31++G**	Ref. [30]
R(1,2)	1.515	1.519	1.516
R(2,3)	1.486	1.489	1.486
R(2,4)	1.523	1.527	1.524
R(4,5)	1.231	1.229	1.233
R(4,6)	1.328	1.328	1.327
A(1,2,3)	109.9(7)	110.2(5)	109.6(6)
A(1,2,4)	111.1(2)	112.3(6)	110.6(1)
A(2,4,5)	119.(5)	117.5(3)	119.3(9)
A(2,4,6)	116.6(7)	117.9(5)	116.3(8)
D(1,2,4,5)	103.0	104.5	69.9
D(1,2,4,6)	75.6	74.5	108.3
D(3,2,4,5)	15.9	14.9	49.7

Table 11. Calculated and experimental geometry parameters of protonated ProNH2 using atom numbering in Scheme 1

Name definition	Bond lengths [Å] and angles [°]	
	UHF/6-31G*	UHF/6-31++G**
R(1,2)	1.535	1.532
R(1,3)	1.546	1.544
R(2,4)	1.525	1.522
R(3,5)	1.524	1.520
R(3,6)	1.518	1.514
R(4,6)	1.528	1.522
R(5,7)	1.231	1.229
R(5,8)	1.331	1.330
A(2,1,3)	104.3(1)	104.2(8)
A(1,2,4)	103.7(1)	103.5(2)
A(1,3,5)	114.3(5)	114.4(2)
A(1,3,6)	104.8(9)	104.9(5)
A(5,3,6)	106.3(6)	106.4(5)
A(2,4,6)	103.4(9)	103.5(3)
A(3,5,7)	118.1(7)	118.1(2)
A(3,5,8)	117.2(9)	117.5(4)
A(7,5,8)	124.5(6)	124.3(2)
A(3,6,4)	108.6(9)	108.5(9)
D(3,1,2,4)	37.8	37.8
D(2,1,3,5)	140.2	140.1
D(2,1,3,6)	24.1	23.8
D(1,2,4,6)	36.3	36.8
D(1,3,5,7)	108.8	109.3
D(1,3,5,8)	70.3	69.9
D(6,3,5,7)	6.4	6.1
D(6,3,5,8)	174.4	174.5
D(1,3,6,4)	1.6	0.9
D(5,3,6,4)	123.1	122.6
D(2,4,6,3)	21.6	22.3

Depending of the type of anionic fragment included in the unit cell of each compound or the presence of the solvent molecules as well the observed different intermolecular hydrogen bonds resulted to a deviation of the geometry of the molecules in solid-state towards the most stable obtained conformer in gas phase. All of the amides studied take the protonated NH_3^+ groups by salts formation with anionic Cl⁻ or hydrogensquarate (*HSq*) and the molecules are in a monoactionic state. Only *ArgNH₂.2HCl* (*L*-argininamidium bis (hydrochloride); $C_6H_{17}N_5O^{2+}.2Cl^-$) [10] and *ArgNH2.2HSq* (*L*-Argininamidium bis (hydrogensquarate); 1-[4- (aminocarbonyl) butyl] guanidinium bis (hydrogensquarate), $C_6H_{17}N_5O^{2+}.2C_4HO_4^-$) [31] have the dicationic form protonated at the amino and guanidinium groups and two anionic molecules realize its neutral form (see Scheme 2). Independently, by the salts formation with hydrochloride or hudrogensquarate, the same electronic structure *i.e.* the protonated amino group and the neutral amide one are observed. However, the different number of H atoms in N^+H_3, amide NH_2 and OH leads to different binding modes with Cl$^-$ and hydrogensquarate moieties. Moreover, in the second cases intermolecular hydrogen bonding between both the anionic parts could be determined as well. These interactions could be highly affected on φ. In the cases of the included solvent molecules or the presence of OH groups in the amino acid side chains like serinamide or tyrosinamide ones lead to their participation in the intermolecular interaction as well. The structure each of protonated amino acid amides could be characterized as well with interactions between OH group and corresponding anionic species in the unit cell. These interactions could be affected on the conformation of the side chins, which deviate the χ_1 values.

Independently, that in the cases of *IleNH₂.HCl*, *MetNH₂.HCl*, *SerNH₂.HCl*, *TyrNH₂.HCl*, *TrpNH₂.HCl*, *ThrNH₂.HCl* and *ValNH₂.HCl* only one Cl$^-$ is refined crystallographically [8,9] in their unit cells, different types of intermolecular interactions are established, resulting to a non-systematically deviations of predicted and experimentally observed dihedral angle values (tables 2-11). The crystal packing in these compounds could be grouping into different patterns according the interactions modes between Cl$^-$ ions and amide NH_2 group.

IleNH₂.HCl ($C_6H_{15}N_2O^+.Cl^-$) crystallized in monoclinic cell setting and $P2_1$ space group with Z = 2 [8]. The unit cell contains two amino acid amide molecules and the interaction mode the bifurcated hydrogen bonding of Cl$^-$ ions to two neighboring amide NH_2 groups translated by one unit cell (Scheme 12). It is most frequency observed situation in the crystal structure of amides [8]. The series of hydrogen bonds of type $N^+H_3...Cl^-$ (3.123, 3.173 Å), $N^+H_3...O=C(NH_2)$ (2.842 Å) and $NH_2...Cl$ (3.285, 3.274 Å) are established (Scheme 12), reflected on the deviation of D(2,3,5,7), D(3,5,7,8), D(3,5,7,9) and D(6,5,7,9) varied within $134.9° - 35.9°$ (Scheme 6 and table 3).

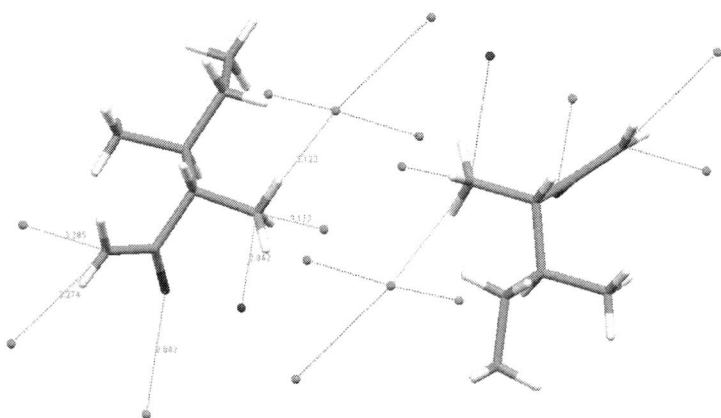

Scheme 12. Intermolecular hydrogen bonds in *IleNH₂.HCl* [8].

TrpNH₂.HCl (2-Ammonio-3-(3-indolyl)propanamide chloride, $C_{11}H_{14}N_3O^+.Cl^-$) and *ValNH₂.HCl* ($C_5H_{13}N_2O^+.Cl^-$) are also with monoclinic cell settings and P2₁ space groups [8]. The interactions with Cl⁻ and neighboring molecules in the unit cells form two types of intermolecular hydrogen bonds, namely NH₂...Cl⁻ and N⁺H₃...Cl⁻ (Schemes 13 and 14) ones. The corresponding bond lengths are 3.270, 3.152, 3.172 Å in *TrpNH₂.HCl* and 3.255 Å; 3.166, 3.210 Å in *ValNH₂.HCl*, respectively. In contrast to *IleNH₂.HCl*, in these cases the C=O group not participates in intra- or intermolecular interactions, which determined the maximal dihedral angle values deviation within 0.0° – 16.3° and less then 5.5° in the case of *TrpNH₂.HCl* (tables 6 and 9) and *ValNH₂.HCl* (tables 6 and 9), respectively. The affectivity of the used level of theory and basis set could be predicted namely using this model system, where the intermolecular or intramolecular interactions in solid phase are refined without the participation of aromatic system in the molecule of *TrpNH₂.HCl* [8]. In last compound the changes in he dihedral angles are experimentally distinguishable because the ab initio and experimental values (table 6) differ by much more the uncertainly, which we estimate to be 2°. Similar tendency is established as well comparing the theoretical predicted geometry parameters of protonated tryptophanamide (table 6) and experimental single crystal X-ray data of other derivatives of tryptophanamide as aqua-(N-acetylhistidinyl-tris(glycyl)tryptophanamide)-copper(ii) pentahydrate [77], N-acetyl-$_{DL}$-tryptophan-N-methylamide [78], threo-□-hydroxy-N-acetyl-tryptophanamide [79] and N□-acetyl-N-methyl-$_L$-tryptophanamide [80]. The obtained bond length and angle differences varies within 0.023 – 0.067 Å, 7.8(9) – 9.9(0)° and 0.0 – 2.6° (in aromatic ring), which illustrated again the applicability of UHF/6-311++G** method for obtaining the adequate structural information.

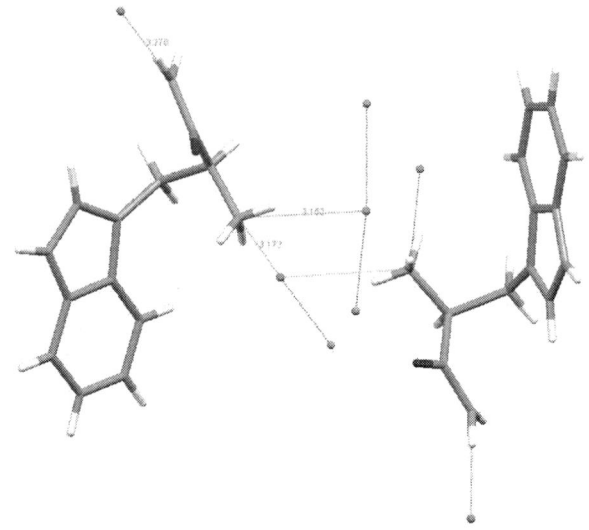

Scheme 13. Intermolecular hydrogen bonds in *TrpNH₂.HCl* [8].

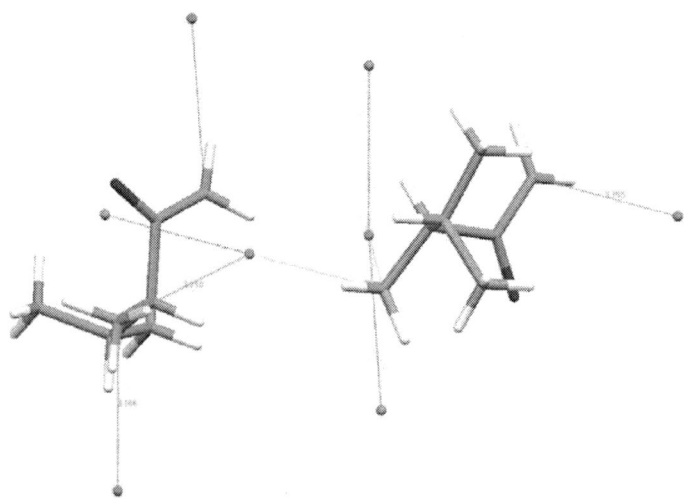

Scheme 14. Intermolecular hydrogen bonds in *ValNH₂.HCl* [8].

In the case of *MetNH₂.HCl* (monoclinic cell setting; P2₁ space group, Z = 2, $C_5H_{13}N_2OS^+.Cl^-$ [8]) only N^+H_3 group and amide O-atom are included in intermolecular interactions (Scheme 15) forming $N^+H_3...Cl^-$ (3.170, 3.234 Å) and $N^+H_3...O=C(NH_2)$ (2.854 Å), respectively. The significant deviation comparing experimental and theoretical data of the φ and χ₁ is obtained as far as the dihedral angle differences are within 135.7° and 28.8° (table 4 and Scheme 7). In this case the influence of dihedral angles of methyonyl side chain is also established.

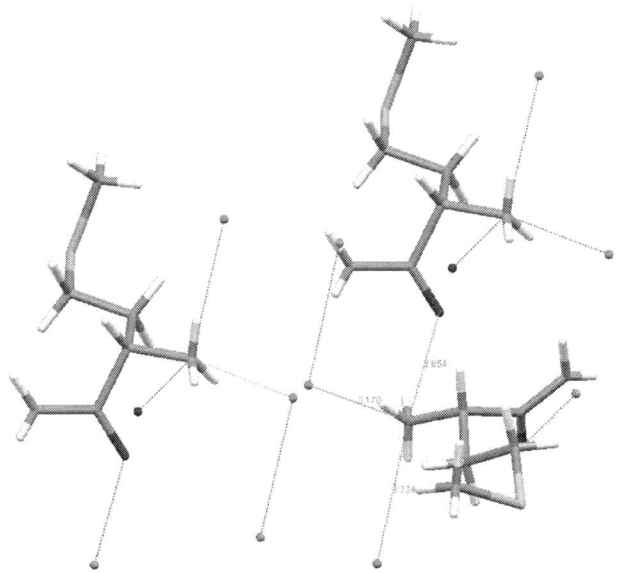

Scheme 15. Intermolecular hydrogen bonds in *MetNH₂.HCl* [8].

In the cases of *SerNH₂.HCl* ($C_3H_9N_2O_2^+.Cl^-$, monoclinic cell setting, P2₁ space group, Z = 2 [8]) and *ThrNH₂.HCl* ($C_4H_{11}N_2O_2^+.Cl^-$, orthorhombic cell setting, P2₁2₁2₁ space group, Z = 4 [8]) the presence of additional protonodonation group as OH and NH₂ resulted to a complicated intermolecular interactions in solid-state (Schemes 16, 17) and resulted to an excepting of significant differences between calculated and experimental observed dihedral angle values. Moreover, in *ThrNH₂.HCl* the unit cell contains four different oriented molecules of the corresponding amino acid amide, which supposed the bigger variety of intermolecular interactions between neighboring *ThrNH₂* molecules (Scheme 17). In *SerNH₂.HCl* are formed $N^+H_3...Cl^-$ (3.169, 3.227Å), $N^+H_3...O=C-NH_2$ (2.783 Å), $NH_2...Cl^-$ (3.259 Å) and $NH_2...OH$ (2.943 Å), respectively (Scheme 16). In *ThrNH₂.HCl* the Cl⁻ are bifurcate hydrogen bonded to two neighboring NH$_2$ groups translated but a died screw symmetry. The corresponding interactions are N$^+$H$_3$...Cl$^-$ (3.155 Å), N$^+$H$_3$...OH (2.965 Å), N$^+$H$_3$...O=C-NH$_2$ (2.898 Å) and NH$_2$...Cl$^-$ (3.279 Å), respectively. The dihedral angle differences in both the cases varies within 97.3° – 9.8° and 112.3° – 6.7°, respectively (Scheme 8, tables 5 and 8). The theoretical predicted geometry parameters of protonated threoninamide vary the level of theory and basis set are presented in table 8. The crystallographic data of *ThrNH$_2$.HCl* included in table 8, indicated that bond lengths and angles are not differ then 0.040 Å, 0.050 Å, 0.070 Å, 1.5°, 2.7° and 6.3°, comparing with experimental data depending of the corresponding theoretical approximations varying level of theory. As can be see the ab initio calculations give more precise bong lengths and angles values, than DFT ones (table 8). The theoretical dihedral angle values depend of the participation of corresponding groups in intermolecular hydrogen bond interactions like in all the presented structures and in this case the C-O bond length of COH group is also influenced due to the formed in solid-state OH...Cl$^-$ and OH...N$^+$H$_3$ intermolecular bonds with lengths of 3.058 Å and 2.965 Å, respectively [8]. The maximal deviation of theoretical and experimental bond length of discussed band is 0.097 Å (MP2/6-311++G*), 0.088 Å (UHF/6-311++G*) and 0.102 Å (B3LYP/6-31++G*). Similar to results of *TyrNH$_2$.HCl* the MP2 and UHF at 6-311++G** basis set gives the similar theoretical results about the geometry parameters with 3.1 % and 2.9 % respectively (table 8).

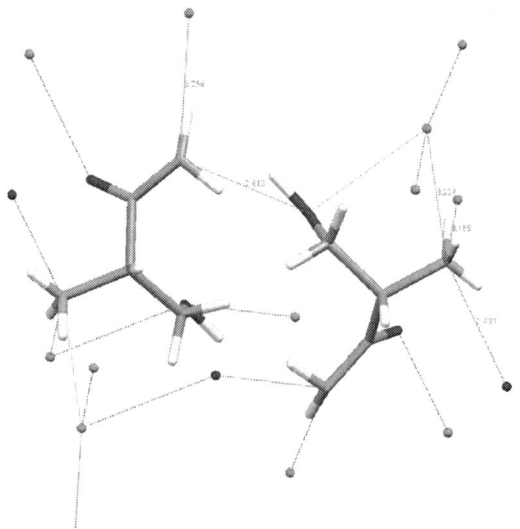

Scheme 16. Intermolecular hydrogen bonds in *SerNH$_2$.HCl* [8].

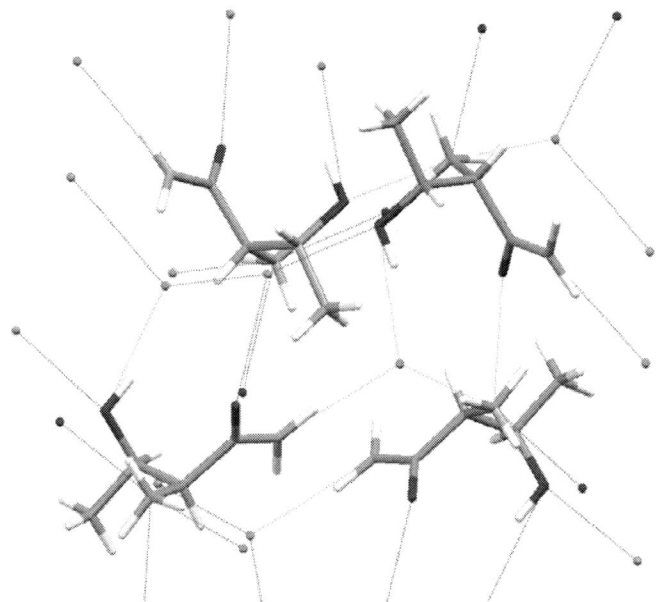

Scheme 17. Intermolecular hydrogen bonds in protonated *ThrNH₂* [8].

In the cases of *ArgNH₂.2HCl* (orthorhombic cell setting, Z = 4, P2₁2₁2₁ [8]) and *TyrNH₂.HCl* (monoclinic cell setting, Z = 2, P2₁ space group [9]) the absence of regularity of the dihedral angle deviations comparing the theoretical results and experimental data could be demonstrated. Moreover for protonated argininamide and tyrosinamide molecules are refined not only hydrochlorides, but hydrogensquarates (*ArgNH₂.2HSq* and *TyrNH₂.HSq*) [8,9]. These data give a direct evidence about the influence of the type of anionic fragment included in the unite cell of these amino acid amide (tables 2 and 7). The effect of the type of anionic fragment and suitable proton donation groups is remarkable in the case of *ArgNH₂.2HSq*, where the guanidyl fragment in addition complicated the hydrogen interaction picture in solid phase (Scheme 18). The N⁺H₃...Cl⁻ (3.197, 3.206 Å)., N⁺H₃...O=C-NH₂ (2.925 Å) and NH₂...Cl⁻ (3.227 Å) intermolecular interactions are obtained [8]. The guanidyl fragment is included in the intermolecular interactions with Cl⁻ ion forming NH₂...Cl⁻ and NH...Cl⁻ bonds with lengths of 3.220, 3.234, 3.226 and 3.223 Å, respectively. The complication is determined as well of the presence of the solvent molecule in the unit cell, resulted to stabilization of NH₂..O and guanydyl NH₂...O bonds with lengths of 2.965 and 2.938 Å, respectively (Scheme 18). These facts affected of the observed deviations of calculated and experimental obtained geometry parameters within

119.7° – 34.5° (table 2 and Scheme 5). The dihedral angles, which name definition contains the corresponding atom included in intermolecular interactions are significant influenced depending of the anionic fragment and solvent molecules in the unit cell. The value differences for D(2,4,5,6) in *ArgNH$_2$.2Cl* and *ArgNH$_2$.2HSq* are 85.5° and 42.0°, respectively. In contrast of D(4,5,6,7), where a significant difference with 119.7° is obtained only in *ArgNH$_2$.2HSq* (table 2). Contravise, the most affected dihedral angle values only in *ArgNH$_2$.2Cl* are D(6,7,8,9), D(7,8,9,11), D(10,8,9,11) and D(10,8,9,12) with 115.6° -12.1°, 18.2° – 16.9°, respectively (Scheme 5 and table 2). As could be see (table 7) the participation of the amide O=C-NH$_2$, and N$^+$H$_3$ groups in intermolecular interactions in *TyrNH$_2$.HCl* resulted to a deviation of dihedral angle values, connected with these fragments varies within 116.6 – 10.7° as far as *TyrNH$_2$.HCl* is characterized with Cl$^-$ and presence of the solvent molecule in the unit cell determining the formation of follow intermolecular hydrogen bonds (Scheme 18): N$^+$H$_3$...O=C-NH$_2$ (2.848 Å), N$^+$H$_3$...OH$_2$ (2.884, 2.890 Å), N$^+$H$_3$...Cl$^-$ (3.220 Å), NH$_2$..OH (2.905), OH...Cl$^-$ (3.058 Å), H$_2$O...Cl$^-$ (3.160, 3.252 Å), respectively [9].

The aromatic fragment is predicted to be planar with maximal deviation of the ring plane of 0.5°, comparing with experimental data where the value is 0.6° (table 7). The changes in dihedral angles of aromatic fragment giving the information of the best-fit values differ of the used basis set and level of theory calculated. The comparison of obtained experimental and theoretical data which we estimated to be ± 2.8 %. These data are similar to UHF/6-311++G**, published in [1,6,7], where the validity of ± 3 % is obtained. As far as the experimental dihedral angles changes of aromatic fragment in both protonated forms studied of tyrosinamide, *TyrNH$_2$.HCl* and *TyrNH$_2$.HSq* [9,76] is ± 3.1% it could be concluded that the both the theoretical methods for the approximation of electronic structure in gas phase are useful to be applied for the calculations. The similar conclusions are obtained comparing the theoretical data (tables 2 and 7) with experimental single crystal X-ray ones of N-acetyl-$_L$-tyrosine-methylamide [81] and N-□-acetyl-$_L$-arginine ethylamide perchlorate [82] giving changes of geometry parameters of ± 3.6 and ± 4.1 %, respectively.

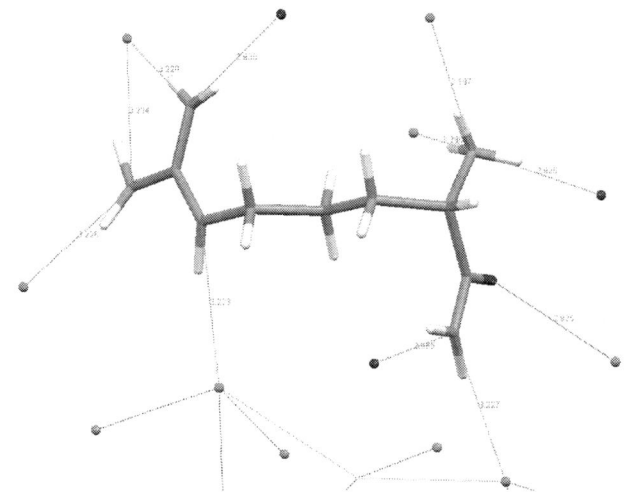

Scheme 18. Intermolecular hydrogen bonds in *ArgNH₂.2HCl* [8].

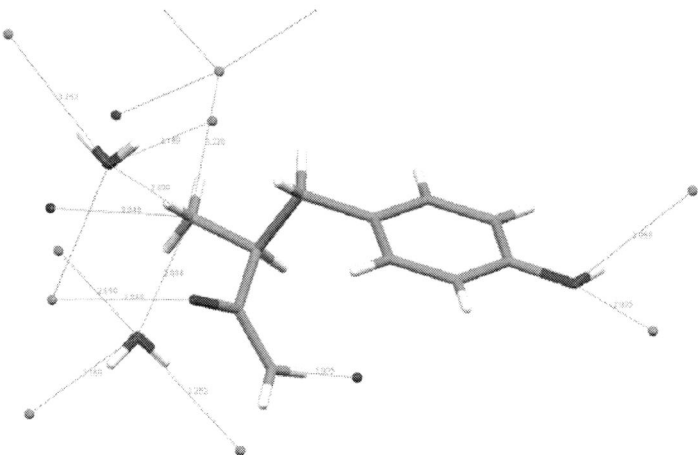

Scheme 19. Intermolecular hydrogen bonds in *TyrNH₂.HCl* [9].

In corresponding hydrogensquarates of tyrosinamide and argininamide the following intermolecular interactions are determined due to the presence of suitable proton donation and accepting fragment as hydrogensquarate and the solvent molecules (*ArgNH₂.2HSq*: $N^+H_3...O=C(Sq)$ (3.007, 2.836, 2.841 Å), $N^+H_3...O=C-NH_2$ (2.831 Å), (Sq)OH...O=C(Sq) (2.565, 2.523 Å) and the following interactions with guanilyl group $NH_2...O=C(Sq)$ (2.842, 2.864, 3.066,

2.897 Å), NH...O=C(Sq) (3.035 Å) ; $TyrNH_2.HSq$: (Tyr)OH...O=C(Sq) (2.727 Å), O=C-NH_2...OH(Tyr) (2.991 Å), O=C-NH_2...O=C-NH_2 (3.068 Å), N^+H_3...O=C(Sq) (2.737, 2.954, 2.607 Å), HOH...OH(Sq) (2.837 Å) and (Sq)OH...OH_2 (2.607 Å) [31], Schemes 20 and 21)

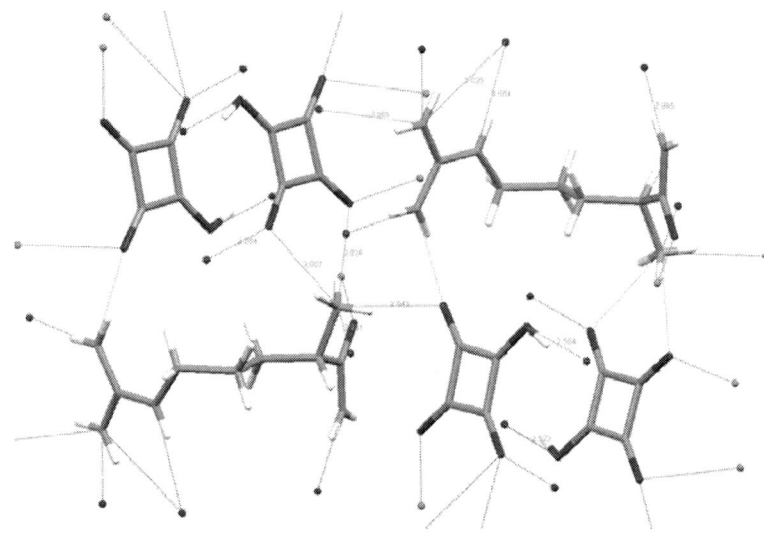

Scheme 20. Intermolecular hydrogen bonds in $ArgNH_2.2HSq$ [31].

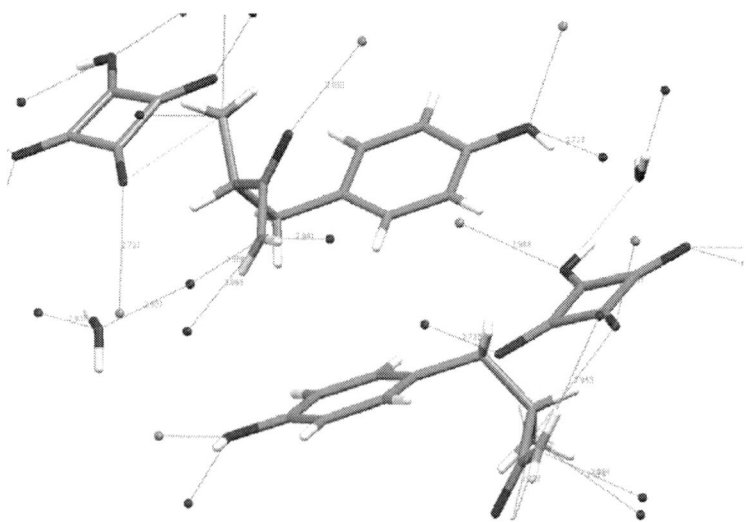

Scheme 21. Intermolecular hydrogen bonding in $TyrNH_2.HSq$ [76].

Results and Discussion

The presented intermolecular interactions in solid-state affected on the complex IR-spectroscopic patterns of the entire compound studied, which detailed interpretation and assignment is difficult for conventional IR techniques. The linear polarized IR-spectroscopic method based on oriented solid samples as a suspension in nematic liquid crystal, combined with reducing-difference procedure and theoretical approximation of electronic structures and vibrational analysis reduces these difficulties to a significant level as has been demonstrated in a series of papers [2-5,10-13]. This is new approach, published for first time in 2004 [83] and is applied in series of organic molecules and complexes as aminopuridines [84-87], heterocyclyc compounds [88-691], polymorphs of glycine, paracetamol and aspirin [92-94], isophorones [95,96], codeine derivatives [97,98] and amino acids, peptides and their salts and Au(III) complexes [28,32,33,74,99-117]. In the cases, where the compounds has been refined with single crystal X-ray diffraction, the validity of the method has been proved and it is appeared as a very comfortable approach for solid-state investigations and vibrational assignment die to its is independent towards the amorphous or crystalline character of the samples. The experimental conditions and procedures for interpretation of the spectra have been validated in [118,119]. The oriented solid samples were obtained as a suspension in a nematic liquid crystals of the 4'-cyano-4'-alkylbicyclohexyl types (ZLI 1695 or MLC 6815, Merck), mesomorphics at room temperature. The relatively weak IR-spectrum of ZLI 1695 permits the recording of the guest-compound bands in the whole 4000 - 400 cm^{-1} range (figure 2). The used for IR-LD analysis IR-spectral ranges of MLC 6815 are 4000 – 3000 cm^{-1}, 2800 – 1750 cm^{-1}, 1700 – 1460 cm^{-1} and 1000 - 400 cm^{-1}, respectively (figure 3). The presence of an isolated nitrile stretching IR-band at 2236 cm^{-1} serves additionally as an orientation indicator. The effective orientation of the samples was achieved through the following procedure: 5 mg of the compound to be studied was mixed with the liquid crystal substance until a slightly viscous suspension was obtained. The phase thus prepared was pressed between two KBr-plates for which, in advance, one direction had been rubbed out by means of fine sandpaper. The grinding of the mull in the rubbing direction promotes an additional orientation of the sample [119].

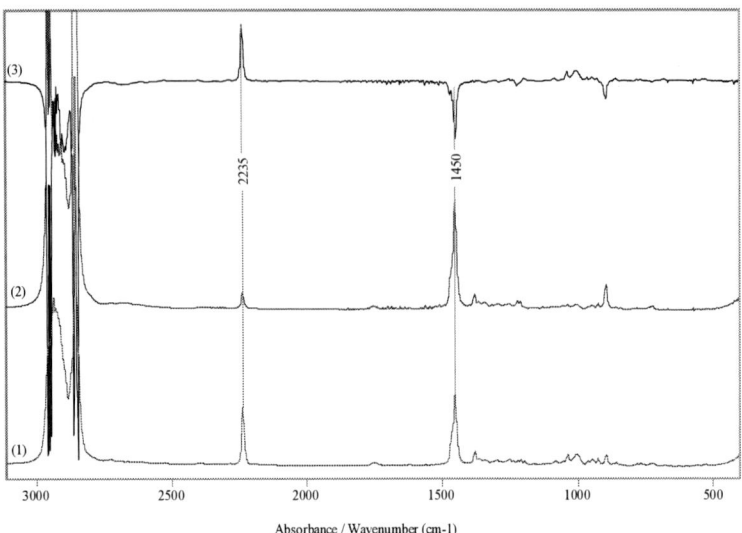

Figure 2. Parallel (1), perpendicular (2) and difference (3) IR-LD spectra of liquid crystal ZLI 1695.

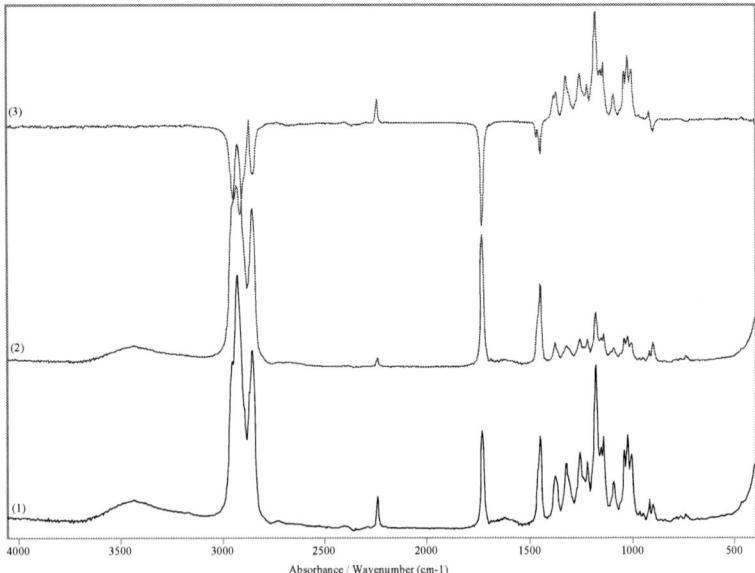

Figure 3. Parallel (1), perpendicular (2) and difference (3) IR-LD spectra of liquid crystal ZLI 6815.

IR-LD spectroscopy background and the interpretation of the linear-polarized IR-spectra are described in [28,32,33,74,105-123]. The method consists of subtraction of the perpendicular spectrum, (IR_s, resulting from a 90^0 angle between the polarized light beam electric vector and the orientation of the sample) (figures 2.2 and 3.2) from the parallel one (IR_p) obtained with a co-linear mutual orientation (figures 2.1 and 3.1). The recorded difference (IR_p-IR_s) spectrum (figures 2.3 and 3.3) divides the corresponding parallel (A_p) and perpendicular (A_s) integrated absorbencies of each band into positive values originating from transition moments, which form average angles with the orientation direction (n) between 0^0 and 54.7^0 (magic angle), and negative ones corresponding to transition moments between 54.7^0 and 90^0 (Scheme 22). In the reducing-difference procedure, the perpendicular spectrum multiplied by the parameter c, is subtracted from the parallel one and c is varied until at least one band or sets of bands are eliminated. The simultaneous disappearance of these bands in the reduced IR-LD spectrum (IR_p - cIR_s) obtained indicates co-linearity of the corresponding transition moments, thus yielding to information regarding the mutual disposition of the molecular fragments. The optimization of the above mentioned procedure for orientation of the solid-samples as well as the used one for IR-spectra interpretation as determination of the position (v_i) and integral absorbencies (A_i) for each i-peak by deconvolution and curve-fitting procedures at 50:50% ratio of Lorentzian to Gaussian peak functions, χ^2 factors within 0.00002–0.00016 and 2000 iterations have been described in [118,119]. The means of two treatments were compared by Student t-test. The experimental IR-spectral patterns have been acquired and processed by GRAMS /AI 7.01 IR spectroscopy (Thermo Galactic, USA) and STATISTICA for Windows 5.0 (StatSoft, Inc., Tulsa, OK, USA) program packages. Accuracy, repeatability, the influence of the liquid crystal medium on the peak positions and integral absorbencies of the guest molecule are presented. A determination of the optimal experimental conditions are obtained as well. The experimental design, determining in a qualitative way the impact of four different input experimental factors (number of scans, the rubbing out of the KBr-pellets, the amount of the compound studied included in the liquid crystal medium and the ration of Lorentzian and Gaussian peak functions in the curve fitting procedure) on the spectroscopic signal for five different frequencies, indicated important specific of the system studied.

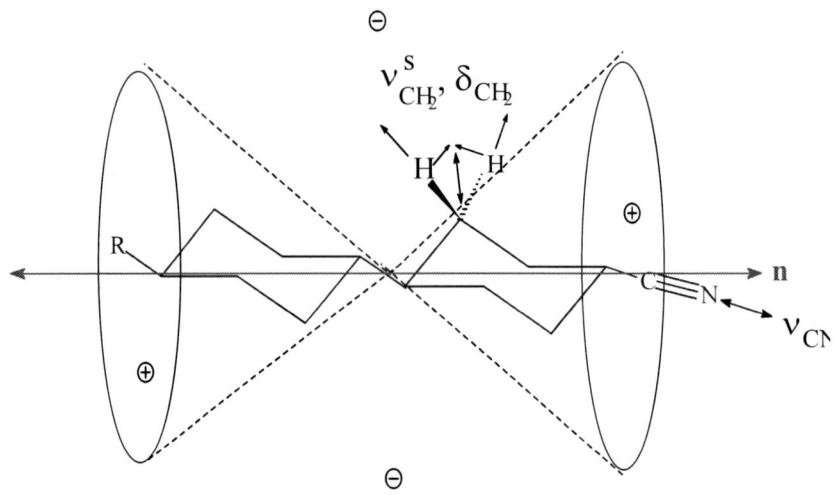

Scheme 22. Some transition moments in ZLI 1695.

The above mentioned complex approach has been demonstrated in series of biological active samples as [28,32,33,74,100-117], where the non-polarized solid-state IR-spectrum shows a significant degree overlapping effect, requesting a preliminary deconvolution, curve-fitting procedure for reasonable IR-bands assignment and experimental IR-LD spectroscopic data, i.e. as may be see from the IR-spectroscopic patterns of *ArgNH$_2$.2HCl*, *MetNH$_2$.HCl*, *SerNH$_2$.HCl* and *TyrNH$_2$.HCl* presented in figure 4. The deconvoluted IR-spectral patterns of *ArgNH$_2$.2HCl* and *MetNH$_2$.HCl* are presented in figures 5, 6.

The complex IR-spectroscopic and theoretical investigation of *TyrNH$_2$.HCl* has been published in [103]. The obtained data indicated that: (i). Series of maxima at 3380 cm^{-1} and 3334 cm^{-1} are assigned as ν_{OH} stretching modes of water molecule and *p*-substituent OH-group in tyrosinamide fragment, respectively; (ii) Peaks at 3376 cm^{-1}, 3210 cm^{-1} and 3160 cm^{-1} to ν^{as}_{NH2} and Fermi resonance spitted ν^{s}_{NH2}, as a result of the asymmetric intermolecular NH$_2$...Cl$^-$ interactions, determined by single crystal X-ray diffraction [8]. (iii) Broad maximum between 3000 – 2800 cm^{-1} is assigned to corresponding asymmetric and symmetric NH$_3^+$-stretchings. (iv) 1651 cm^{-1} (δ_{NH2}), 1641 cm^{-1} ($\nu_{C=O}$), 1621 cm^{-1} (δ^{as}_{NH3+}), 1612 cm^{-1} ($\delta^{'as}_{NH3+}$), 1594 cm^{-1} (8a, phenyl radial vibration) 1513 cm^{-1} (δ^{s}_{NH3+}) and 1504 cm^{-1} (19a, phenyl radial vibration). (v) 852 cm^{-1} corresponds to out-of-plane (o.p.) mode of p-disubstituted tyrosyl-fragment [124].

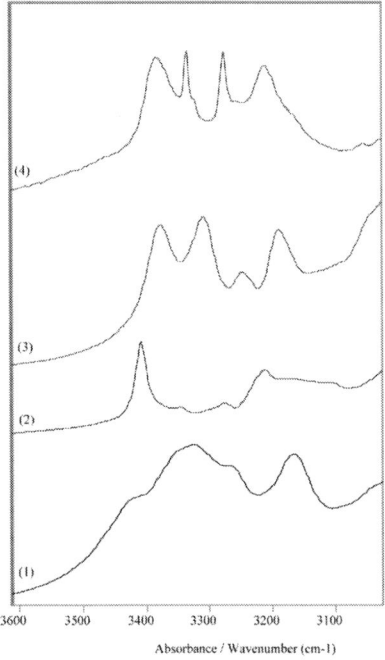
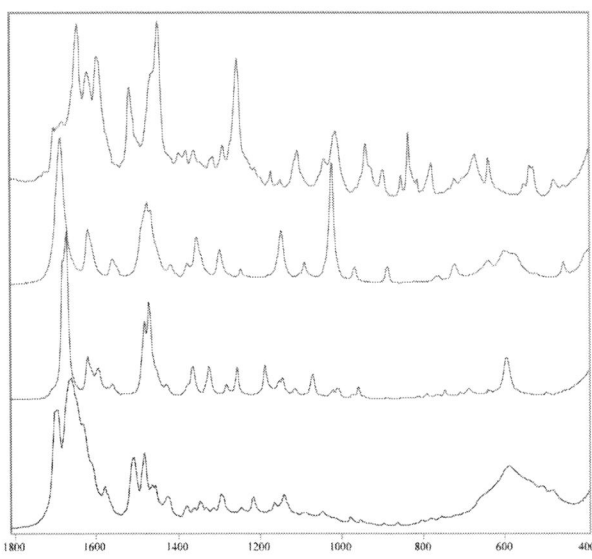

Figure 4. IR-spectra of ArgNH$_2$.2HCl (1), MetNH$_2$.HCl (2), SerNH$_2$.HCl (3) and TyrNH$_2$.HCl (4).

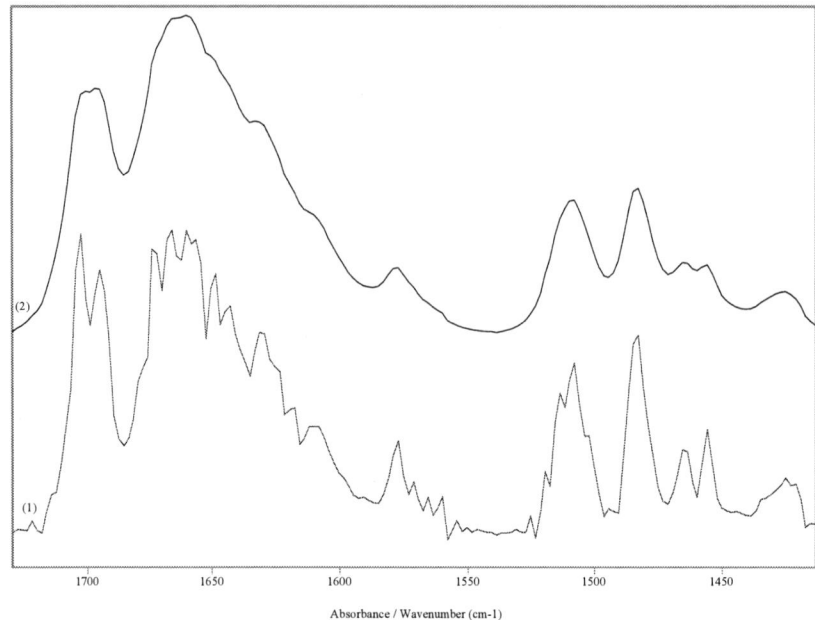

Figure 5. 1700 – 1400 cm^{-1} deconvoluted (1) and not procedure IR-spectrum of *ArgNH$_2$.2HCl* (γ factor 4.1, increment 0.1 and Bessel function).

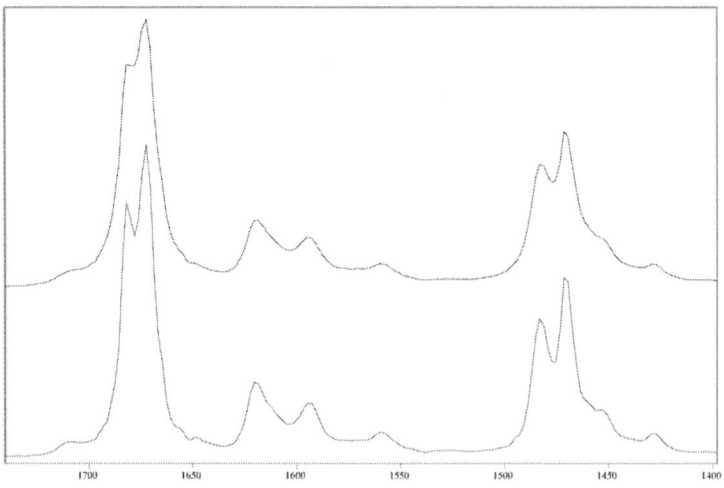

Figure 6. 1700 – 1400 cm^{-1} deconvoluted (1) and not procedure IR-spectrum of *MetNH$_2$.HCl* (γ factor 1.9, increment 0.1 and Bessel function).

Results and Discussion

The IR-characteristic bands assignment of *ArgNH$_2$.2HCl*, *MetNH$_2$.HCl* and *SerNH$_2$.HCl*, obtained by the above mentioned approach is summarized in table 12 and are supposed with known ones about zwitterionic and protonated forms of corresponding amino-acids as $_L$-arginine hydrochloride monohydrate [125], $_L$-argininium dinitrate [126], $_L$-Methionine and its N-deuterated derivative [127] or methionyl containing metal complexes [128-133] obtained for other IR- and Raman spectroscopic studies, thus shown a good correlation is established between our conclusions and published data.

**Table 12. IR-characteristic bands assignment of
ArgNH$_2$.2HCl, *MetNH$_2$.HCl* and *SerNH$_2$.HCl***

Assignment	v [cm^{-1}]		
	ArgNH$_2$.2HCl	MetNH$_2$.HCl	SerNH$_2$.HCl
v^{as}_{NH2}, v^{s}_{NH2}	3322, 3262, 3168	3407, 3212	3308, 3247, 3188
v^{as}_{N+H3}, v^{s}_{N+H3}	3300 – 2600 broad maximum		
v_{OH}	-	-	3375
$v_{C=O}$, Amide I	1696	1674	1687
δ^{as}_{N+H3}, δ^{s}_{N+H3}	1677, 1629, 1508	1681, 1594, 1559	1697, 1606, 1557
δ_{NH2} (Amide II)	1608	1618	1620
v_{C-N} (Amide III)	1381	1363	1377
$\delta_{C=O}$ (Amide IV)	594	595	601
ϖ_{NH2} (Amide V)	661	685	639
τ_{NH2} (Amide VII)	754	765	761

In all the hydrochlorides studied the participation of amide C=O group in intermolecular hydrogen bond interactions (see above) resulted to an observation of corresponding $v_{C=O}$ mode in 1700 – 1680 cm^{-1} (table 12) depending of the power of the interaction. A significant high frequency shifting of δ^{as}_{N+H3}, δ^{s}_{N+H3} as well as of δ_{NH2}, which regions are observed within 1700 – 1500 cm^{-1} IR-spectroscopic region (table 12) and about 1600 cm^{-1} is also explained by the participation of amide NH$_2$ and N$^+$H$_3$ groups in intermolecular hydrogen bonds in solid phase.

The IR-spectral patterns are significant complexity in addition in the cases of hydrogensquarates of amino acid amides, where in parallel are observed the complex IR-spectral bands of hydrogensquarate fragment of the molecule as can be seen from figure 7, presented the IR-spectrum of *ArgNH$_2$.2HSq*.

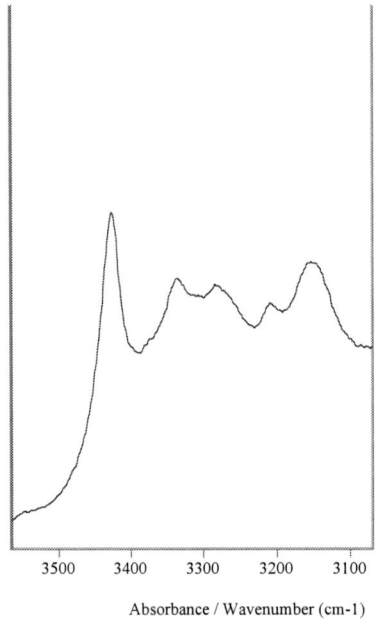

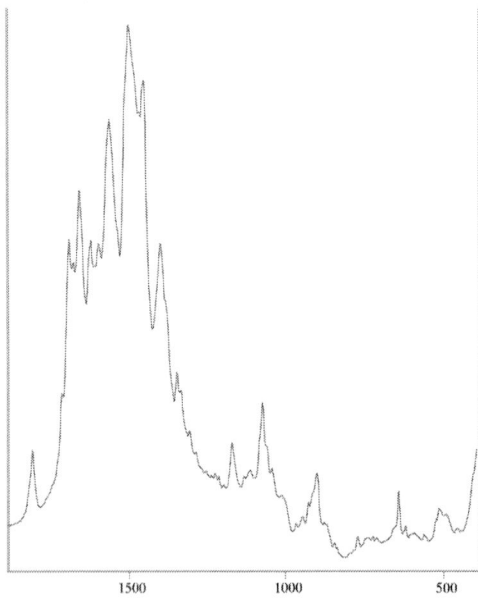

Fig.ure 7. Solid-state IR-spectrum of *ArgNH₂.2HSq* in 4000 – 400 cm⁻¹ IR-spectral region (KBr pellet).

The deconvolution (figure 8) and curve fitting procedures (figure 9) applied for obtaining of the number of peaks and their positions resulted to the observation of series of strong overlapped maxima both in NH- and C=O stretching regions (figures 8 and 9). The data in figure 9 are proposed as follow: 3427 cm^{-1} (v^{as}_{NH2}), 3341 cm^{-1} (v^{as}_{N+H2}), 3273 cm^{-1} (v^{s}_{NH2}), 3209 cm^{-1} (v^{s}_{N+H2}), 3153 cm^{-1} (v_{N+H}), 1805 cm^{-1} ($v^{as}_{C=O(Sq)}$), 1712 cm^{-1} (δ_{N+H2}), 1691 cm^{-1} (δ^{as}_{N+H3}), 1686 cm^{-1} ($v^{s}_{C=O(Sq)}$), 1674 cm^{-1} ($v_{C=O}$, Amide I), 1656 cm^{-1} (δ_{NH2}, Amide II), 1645 cm^{-1} ($\delta^{as'}_{N+H3}$), 1607 cm^{-1} (δ^{s}_{N+H3}) and 1596 cm^{-1} (δ_{N+H}), respectively.

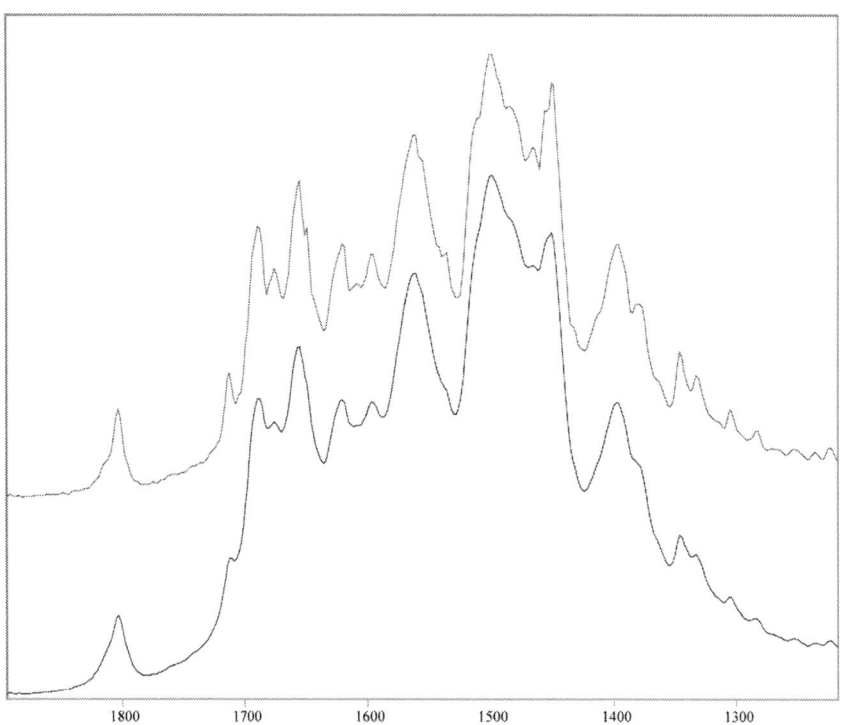

Figure 8. Non- and deconvoluted (γ factor = 3.1 and increment of 0.1) IR-spectral patterns of the solid-state IR-spectrum of $ArgNH_2.2HSq$.

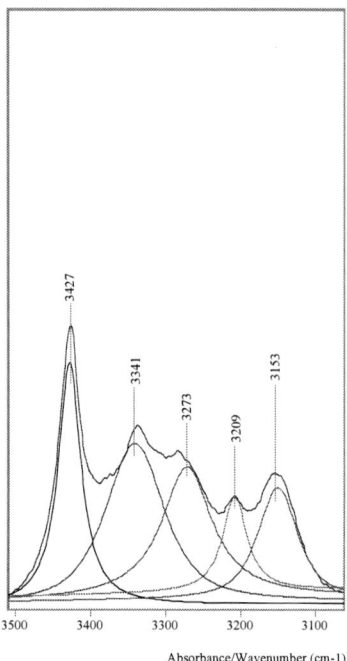
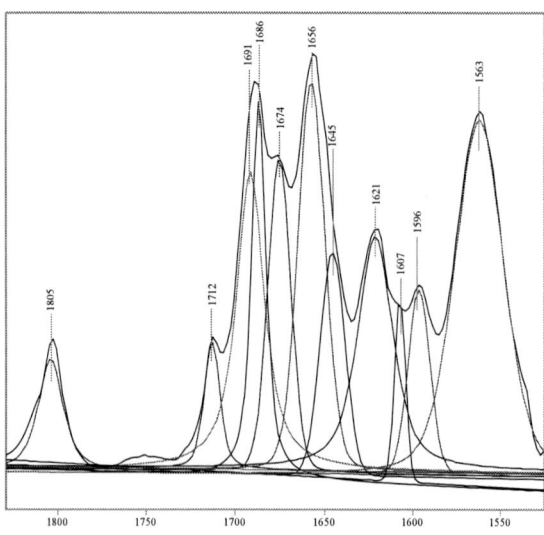

Figure 9. Curve fitted IR-spectral patterns of the solid-state IR-spectrum of $ArgNH_2.2HSq$.

These data correlated well with the experimental single crystal X-ray ones [31], indicating the stronger intermolecular interaction with participation of guanidyl group in contrast to amide NH_2 one, which affected on the lower frequency shifting of symmetric and asymmetric stretching vibrations and higher frequency shifting of corresponding bending modes of protonated guanudyl group. The NH_2 amide fragment is characterized with higher frequency disposed stretching asymmetric peak and lower-frequency one of corresponding bending mode (Amide II). The participation of amide C=O group in intermolecular hydrogen bond (see above) determined the relatively low frequency shifted Amide I peak to 1674 cm^{-1}, like the typical values for primary amides are within 1680 ± 35 cm^{-1} [124]. These data are confirmed by the obtained polarized IR-spectroscopic data and the made reducing-difference procedure, where the consequently elimination of the peaks at 3273 cm^{-1} (v^s_{NH2}) and 3209 cm^{-1} (v^s_{N+H2}) resulted to disappearance of the maxima at 1656 cm^{-1} (δ_{NH2}, Amide II) and 1712 cm^{-1} (δ_{N+H2}), confirming the origin of the discussed bands as far as in the frame of one molecule the corresponding stretching and bending modes are co-linear oriented (Scheme 23).

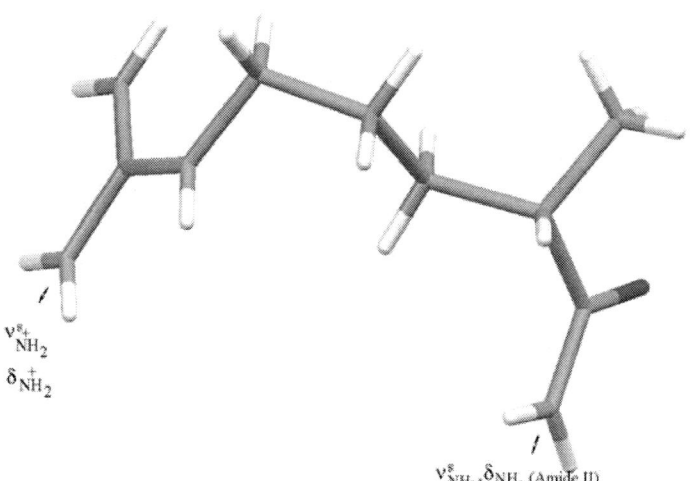

Scheme 23. Some transition moments in protonated *ArgNH₂*.

The intermolecular interactions in solid state resulted to a deviation of corresponding peak positions in experimental and theoretical predicted modes. In the case of protonated argininamide, the calculation of vibrational frequencies and infrared intensities were checked for which kind of calculations performed agree best with the experimental data for the non-affected of intermolecular interactions groups in the molecule. In this case the DFT method provide more accurate vibrational data, as far as the calculated standard deviations 9 cm^{-1} (B3LYP), 22 cm^{-1} (UHF) and 21 cm^{-1}(UMP2), respectively. So, the B3LYP/6-31++G** data are presented for above discussed modes of amide and gaunydyl fragment in protonated arininamide, where for the better correspondence between the experimental and theoretical values, a modification of the results using the empirical scaling factor (0.9614) [134] is made. The theoretical predicted values are 3477 cm^{-1} (v^{as}_{NH2}), 3381 cm^{-1} (v^{as}_{N+H2}), 3300 cm^{-1} (v^{s}_{NH2}), 3236 cm^{-1} (v^{s}_{N+H2}), 3180 cm^{-1} (v_{N+H}), 1825 cm^{-1} ($v^{as}_{C=O(Sq)}$), 1736 cm^{-1} (δ_{N+H2}), 1700 cm^{-1} (δ^{as}_{N+H3}), 1693 cm^{-1} ($v^{s}_{C=O(Sq)}$), 1690 cm^{-1} ($v_{C=O}$, Amide I), 1677 cm^{-1} (δ_{NH2}, Amide II), 1661 cm^{-1} ($\delta^{as'}_{N+H3}$), 1636 cm^{-1} (δ^{s}_{N+H3}) and 1600 cm^{-1} (δ_{N+H}), respectively. As could be see the differences between the theoretical and experimental frequency values between 25-13 cm^{-1} is established, due to the participation of all discussed fragments in intermolecular interactions. In contrast to non affected groups, which theoretical and experimental frequencies are not differ than 7 cm^{-1}, which is in the frame of the accuracy of the used theoretical method.

AMINO ACID AMIDES ESTER AMIDES OF SQUARIC ACID

The theoretical approximation of the geometries of *MetNHSqEt*, *PheNHSqEt* and *ValNHSqEt* at RHF/6-31++G** have been published in [28,32,135]. The results, obtained by MP2 level of theory in same basis set are presented in tables 13-15. The quantum chemical calculations are obtained by the procedure described above. In all cases the predicted geometry parameters as bond lengths and angles are compared with experimental ones, obtained by single crystal X-ray diffraction [23,27,28]. The data included in tables 13-15 corresponds for most stable conformers of *MetNHSqEt*, *PheNHSqEt* and *ValNHSqEt* with E$_{rel}$ 1.2 (C$_4$ in Scheme 24), 0.0 (C$_5$ in Scheme 26) and 0.1 (C$_1$ in Scheme 27), respectively. It is interesting to obtained that for *MetNHSqEt* and *ValNHSqEt* the geometries of the most stable form obtaining by both the theoretical ab initio approximations are similar with dihedral angle differences that do not bigger than 0.1°, respectively (Schemes 24 and 27, tables 13 and 15). The conformational analysis of *MetNHSqEt* predicted 28 conformers and only six with E$_{rel}$ lower than 12 kJ/mol, presented in scheme. The corresponding φ, χ$_1$, and E values of these forms are shown in figure 10.

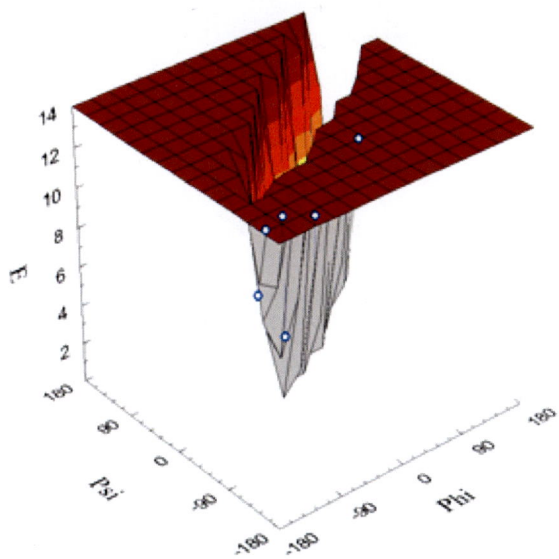

Figure 10. Landscape and contour representation of the φ and χ potential energy surphace of the molecule of *MetNHSqEt* after optimization at RHF/6-311++G**.

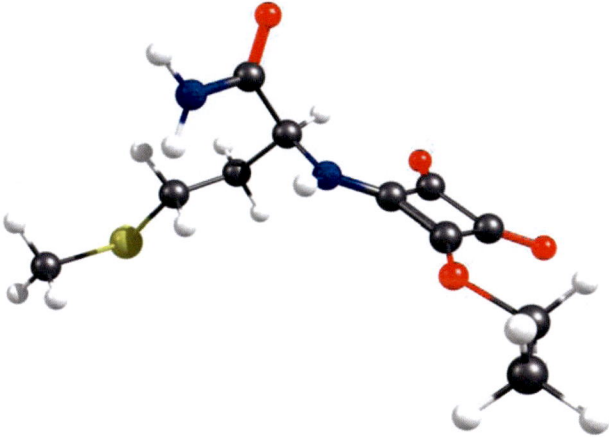

C_1 (E_{rel} = 11.5 kcal/mol).

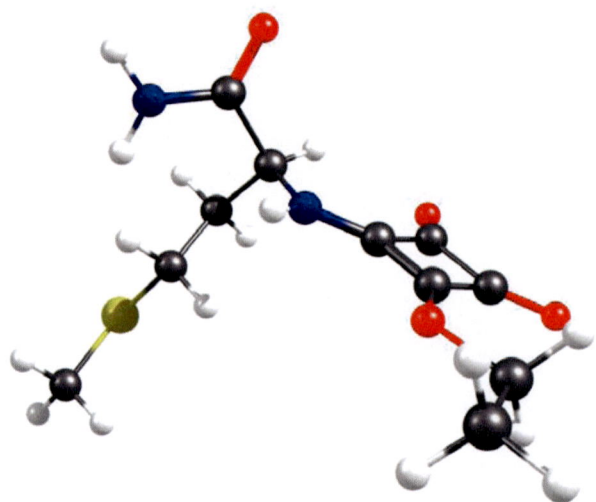

C_2 (E_{rel} = 7.2 kcal/mol).

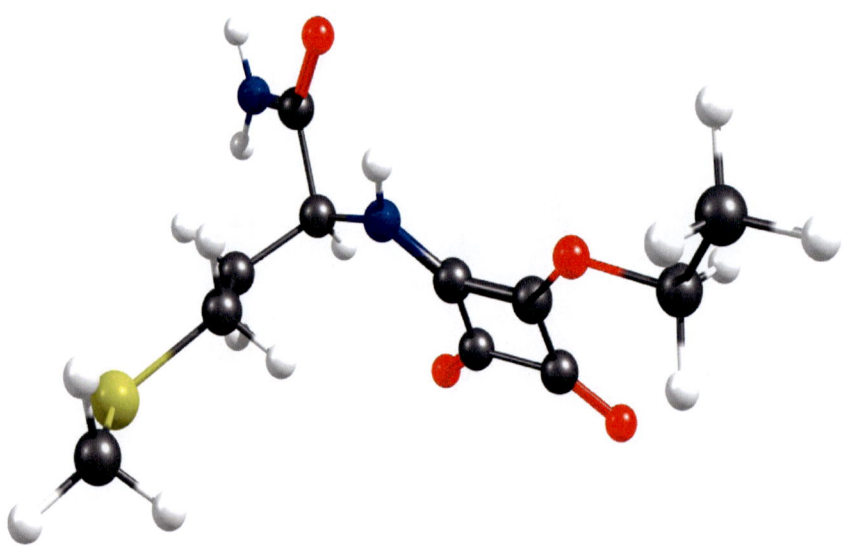

C_3 (E_{rel} = 9.2 kcal/mol).

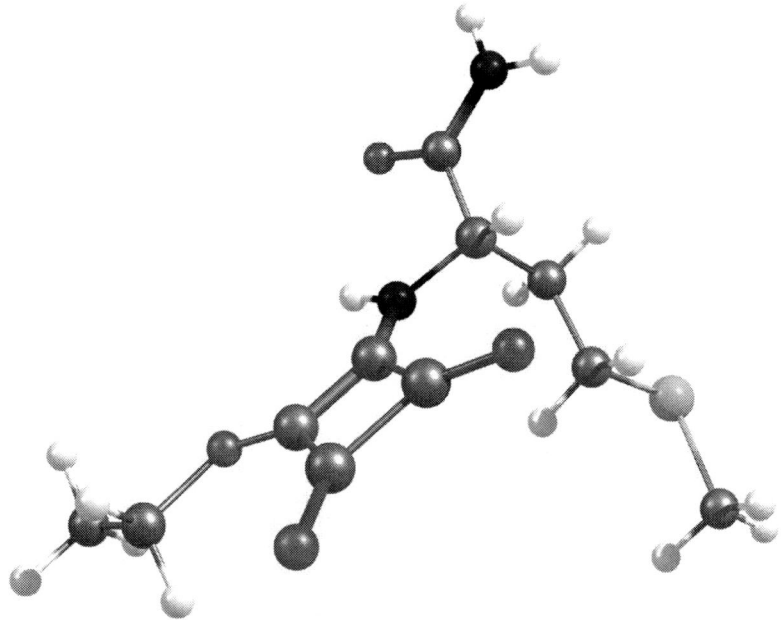

C_4 (E_{rel} = 1.2 kcal/mol).

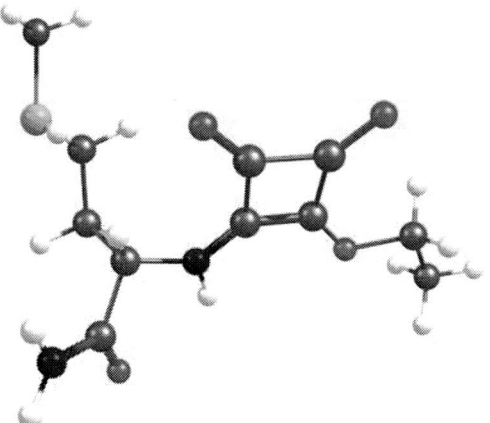

C_5 (E_{rel} = 6.8 kcal/mol).

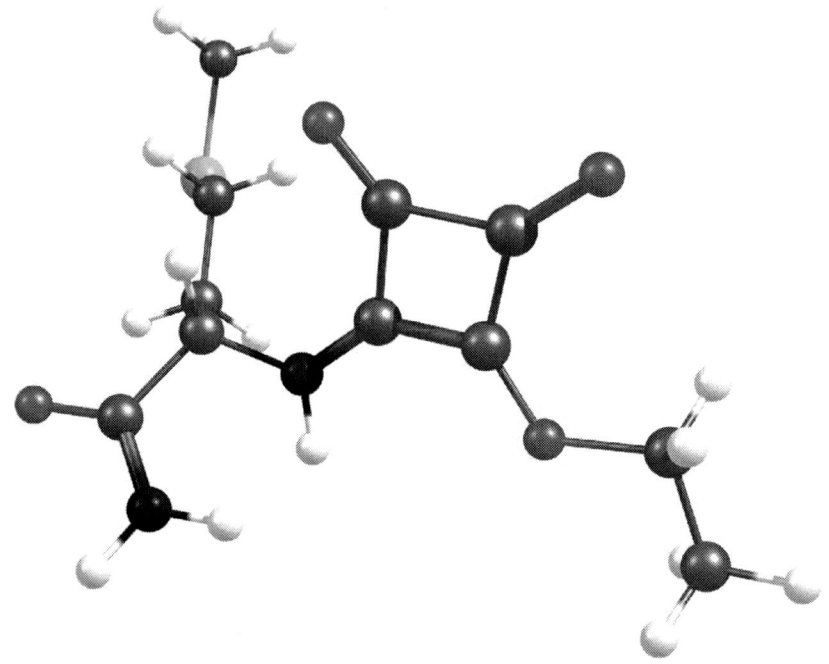

C_6 (E_{rel} = 8.7 kcal/mol).

Scheme 24. Ab initio model conformers (Ci, i = 1-6) of MetNHSqEt after optimization at RHF/6-311++G** level of theory.

In *PheNHSqEt*, the most stable forms with E_{rel} equal of 0.2 kJ/mol (RHF/6-31++G**) and 0.7 kJ/mol (MP2/6-31++G**) are characterized with different geometries in gas phase (table 14) [135]. Comparing with experimental single crystal X-ray diffraction data, the better theoretical approximation with ± 1.8° both for bond lengths and angles of aromatic fragment in this case is using RHF level of theory. The geometry obtained using MP2 approximation, indicated the stabilization of NH_2...O=C(Sq) intramolecular hydrogen bond with length of 3.000 Å (Scheme 25). In this case from the 29 predicted potential energy minima, only eight are characterized with E_{rel} lower than 12 kJ/mol (Scheme 26) and figure 11.

Table 13. Theoretical and experimental geometry parameters of *MetNHSqEt* using atom numbering Scheme 1

Name definition	Bond lengths [Å] and angles [°]	
	HF/6-311++G**	Ref. [27]
R(1,2)	1.769	1.785
R(2,3)	1.798	1.796
R(3,4)	1.505	1.507
R(4,5)	1.545	1.544
R(5,6)	1.512	1.508
R(5,7)	1.454	1.460
R(6,9)	1.227	1.234
R(6,11)	1.306	1.309
R(7,8)	1.332	1.318
R(8,10)	1.392	1.397
R(8,12)	1.484	1.472
R(10,13)	1.332	1.318
R(10,14)	1.464	1.462
R(12,14)	1.504	1.504
R(12,15)	1.209	1.200
R(13,16)	1.463	1.475
R(14,17)	1.211	1.221
A(1,2,3)	99.9(3)	100.6(0)
A(2,3,4)	110.2(9)	113.8(1)
A(3,4,5)	112.3(5)	112.3(7)
A(4,5,6)	108.8(63)	109.4(1)
A(4,5,7)	112.6(1)	110.0(5)
A(5,6,9)	120.5(4)	120.3(2)
A(5,6,11)	116.5(4)	117.8(3)
A(5,7,8)	122.1(9)	122.7(3)
A(7,8,10)	134.9(9)	133.4(3)
A(10,8,12)	91.6(0)	90.8(9)
A(8,10,13)	130.1(5)	128.5(6)
A(8,10,14)	94.2(4)	93.2(7)
A(8,12,14)	87.5(2)	88.6(5)
A(8,12,15)	134.8(2)	136.2(5)
A(10,13,16)	120.3(2)	117.2(3)
D(1,2,3,4)	179.9	71.9
D(2,3,4,5)	175.5	177.6
D(3,4,5,6)	177.3	177.3
D(3,4,5,7)	63.5	61.6
D(4,5,6,9)	93.6	83.5
D(4,5,6,11)	84.1	94.3
D(7,5,6,9)	28.6	37.5
D(7,5,6,11)	153.6	144.6
D(4,5,7,8)	103.5	127.4
D(5,7,8,10)	177.3	178.6
D(7,8,10,13)	0.4	1.8
D(7,8,10,14)	179.9	179.4
D(12,8,10,14)	0.1	0.9
D(7,8,12,15)	0.1254	2.8
D(10,8,12,15)	179.9	175.4
D(8,10,13,16)	177.6	175.9
D(8,10,14,17)	179.9	179.5
D(15,12,14,17)	0.1	2.6
D(10,13,16,18)	179.6	100.2

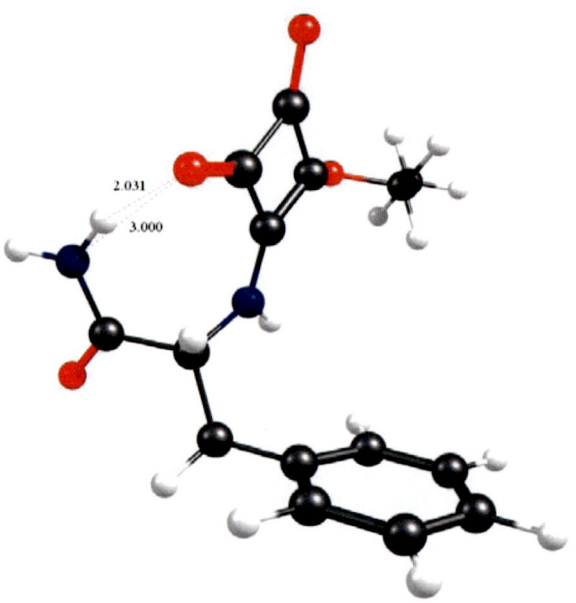

Scheme 25. Ab initio model conformer of *PheNHSqEt* after optimization at MP2/6-311++G** level of theory [135].

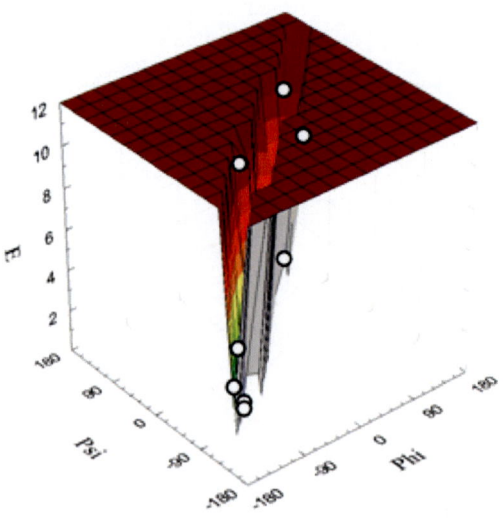

Figure 11. Landscape and contour representation of the φ and χ potential energy surphace of the molecule of *PheNHSqEt* after optimization at RHF/6-311++G** [135].

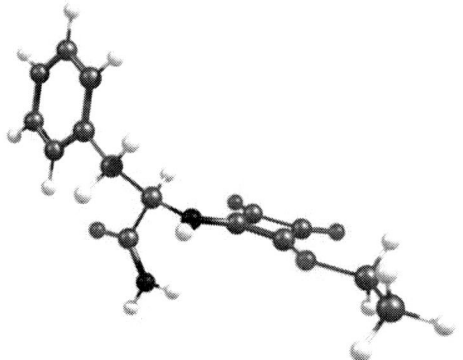

C_1 (E_{rel} = 6.1 kcal/mol).

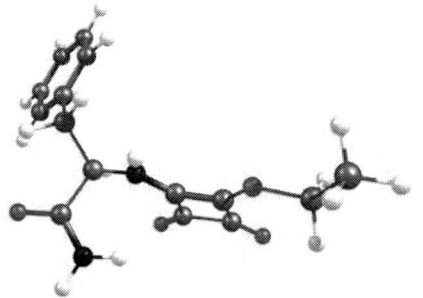

C_2 (E_{rel} = 3.2 kcal/mol).

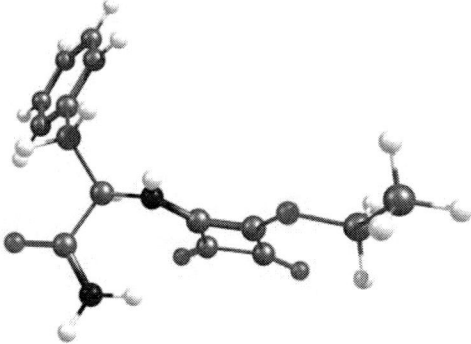

C_3 (E_{rel} = 1.1 kcal/mol).

Scheme 26. (Continued on next page.)

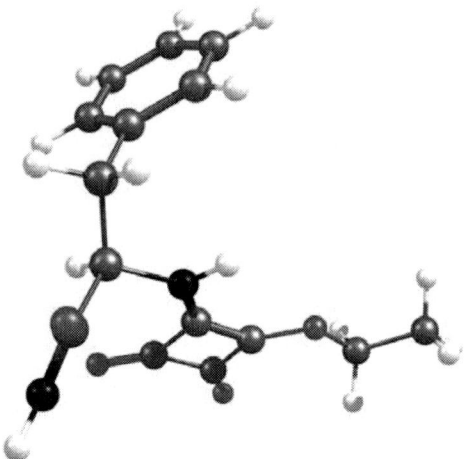

C_4 (E_{rel} = 11.8 kcal/mol).

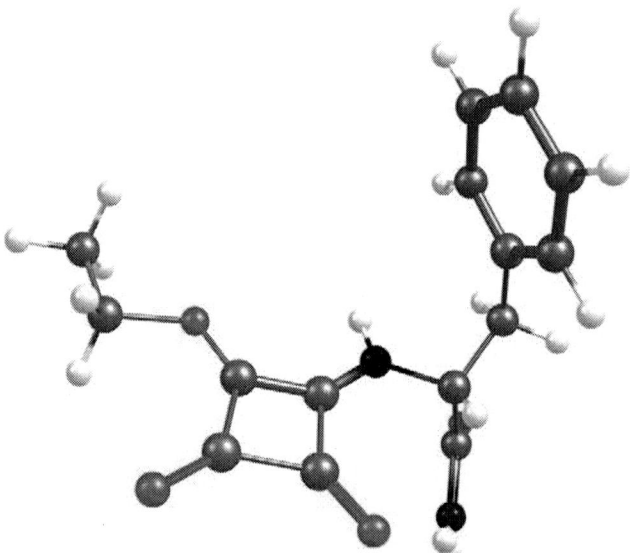

C_5 (E_{rel} = 0.0 kcal/mol).

Scheme 26. (Continued on next page.)

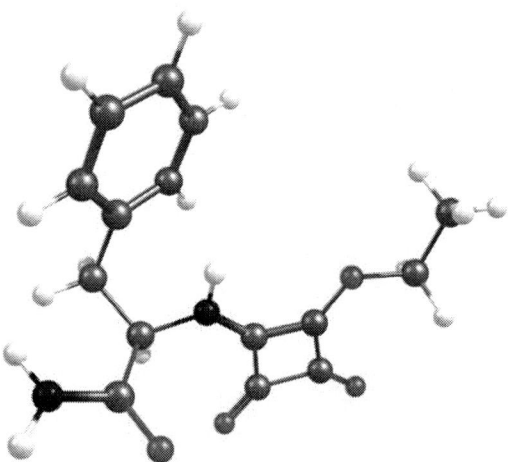

C_6 (E_{rel} = 4.1 kcal/mol).

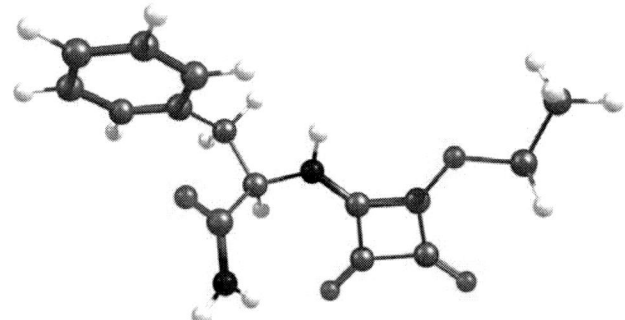

C_7 (E_{rel} = 10.3 kcal/mol).

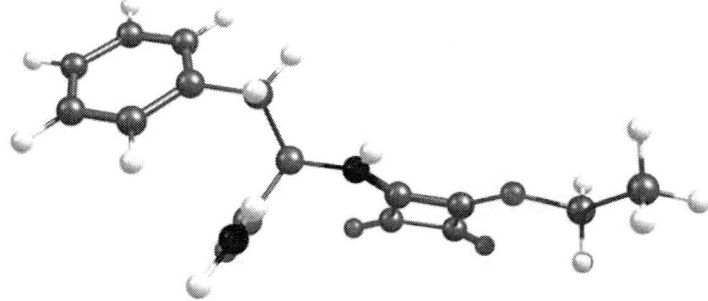

C_8 (E_{rel} = 7.1 kcal/mol).

Scheme 26. Ab initio model conformers (C_i, i = 1-8) of *PheNHSqEt* after optimization at RHF/6-311++G** level of theory.

Table 14. Theoretical and experimental geometry parameters of *PheNHSqEt* using atom numbering Scheme 1

Name definition	Bond lengths [A] and angles [°]		
	UHF/6-311++G**[135]	MP2/6-31++G**[135]	Ref. [23]
R(1,2)	1.385	1.396	1.374
R(1,6)	1.393	1.400	1.377
R(2,3)	1.388	1.395	1.345
R(3,4)	1.385	1.396	1.380
R(4,5)	1.388	1.396	1.387
R(5,6)	1.390	1.399	1.389
R(6,7)	1.512	1.512	1.488
R(7,8)	1.534	1.538	1.536
R(8,9)	1.528	1.523	1.531
R(8,10)	1.465	1.479	1.456
R(9,11)	1.345	1.359	1.301
R(9,12)	1.228	1.206	1.227
R(10,13)	1.333	1.372	1.316
R(13,14)	1.372	1.378	1.385
R(13,16)	1.480	1.495	1.467
R(14,15)	1.330	1.356	1.316
R(14,18)	1.465	1.485	1.441
R(15,17)	1.464	1.414	1.450
R(16,18)	1.521	1.491	1.493
R(16,20)	1.212	1.219	1.223
R(17,19)	1.509	1.525	1.217
R(18,21)	1.209	1.221	1.218
A(2,1,6)	120.8(6)	120.9(7)	121.0(2)
A(1,2,3)	120.1(9)	119.9(4)	121.1(3)
A(2,3,4)	119.5(4)	119.6(3)	118.9(6)
A(3,4,5)	120.1(1)	120.1(2)	121.1(5)
A(4,5,6)	120.895	120.7882	119.2(4)
A(1,6,5)	118.4(4)	118.5(0)	118.4(8)
A(1,6,7)	120.7(1)	120.3(2)	121.4(1)
A(6,7,8)	113.3(1)	112.4(0)	113.9(9)
A(7,8,9)	110.7(9)	110.6(9)	108.7(7)
A(7,8,10)	110.6(9)	109.7(0)	110.9(6)
A(8,9,11)	115.6(0)	114.7(1)	117.8(5)
A(11,9,12)	123.3(0)	124.7(3)	123.7(8)
A(8,10,13)	123.6(9)	118.7(2)	124.6(4)
A(10,13,14)	135.3(8)	133.7(2)	137.0(9)
A(10,13,16)	83.0(0)	92.0(9)	88.6(7)
A(14,13,16)	91.6(0)	92.0(9)	90.0(0)
A(13,14,15)	130.1(7)	130.7(9)	129.5(3)
A(13,14,18)	94.1(8)	92.1(7)	94.2(2)
A(14,15,17)	120.3(5)	113.8(9)	116.1(2)
A(13,16,20)	134.6(1)	135.6(6)	134.9(1)

Name definition	Bond lengths [Å] and angles [°]		
	UHF/6-311++G**[135]	MP2/6-31++G**[135]	Ref. [23]
A(15,17,19)	106.4(9)	108.8(1)	115.3(7)
A(14,18,21)	136.9(8)	135.5(6)	137.0(4)
D(6,1,2,3)	0.2	0.1	0.8
D(2,1,6,5)	0.6	0.2	0.9
D(2,1,6,7)	179.7	179.9	175.8
D(1,2,3,4)	0.2	0.1	0.3
D(2,3,4,5)	0.2	0.1	0.1
D(3,4,5,6)	0.1	0.0	0.3
D(4,5,6,1)	0.5	0.2	0.7
D(1,6,7,8)	70.5	88.3	88.3
D(6,7,8,9)	178.8	170.2	174.7
D(6,7,8,10)	60.4	68.5	64.6
D(7,8,9,11)	146.6	166.2	114.7
D(10,8,9,11)	91.4	72.0	124.7
D(10,8,9,12)	87.0	108.3	105.0
D(8,10,13,14)	175.5	177.6	8.1
D(10,13,14,15)	0.3	16.4	1.7
D(10,13,14,18)	179.7	16.4	178.7
D(16,13,14,15)	178.6	160.3	179.7
D(10,13,16,20)	2.1662	2.5778	1.83
D(14,13,16,20)	179.8	179.3	179.4
D(13,14,15,17)	179.9	107.7	174.1
D(13,14,18,21)	179.1	175.9	179.8
D(14,15,17,19)	177.9	178.9	177.4
D(13,16,18,14)	1.1	2.8	0.6
D(13,16,18,21)	179.2	176.1	179.8

The corresponding conformational data of *ValNHSqEt* are presented in Scheme 27 and figure 12, where is shown the seven conformers with potential energy minima lower than 12 kcal/mol, from whose the most stable one corresponds to E_{rel} equal of 0.2 kcal/mol (C_1 in Scheme 27).

Table 15. Theoretical and experimental geometry parameters of *ValNHSqEt* using atom numbering Scheme 1

Name definition	Bond lengths [Å] and angles [°]	
	HF/6-311++G**	Ref. [28]
R(1,2)	1.538	1.494
R(2,3)	1.536	1.498
R(2,4)	1.555	1.551
R(4,5)	1.467	1.448
R(4,6)	1.533	1.500
R(5,9)	1.343	1.313

Table 15. (Continued).

Name definition	Bond lengths [Å] and angles [°]	
	HF/6-311++G**	Ref. [28]
R(6,7)	1.349	1.291
R(6,8)	1.225	1.256
R(9,10)	1.366	1.393
R(9,11)	1.490	1.470
R(10,12)	1.467	1.474
R(10,13)	1.235	1.288
R(11,12)	1.526	1.490
R(11,15)	1.204	1.220
R(12,16)	1.208	1.213
R(13,14)	1.464	1.475
R(14,17)	1.495	1.394
A(1,2,3)	109.1(5)	111.7(5)
A(1,2,4)	110.8(6)	113.0(2)
A(2,4,5)	110.6(3)	109.1(2)
A(2,4,6)	113.1(6)	111.0(4)
A(4,5,9)	125.9(2)	123.9(1)
A(4,6,7)	116.5(7)	119.1(3)
A(7,6,8)	121.5(7)	121.4(3)
A(5,9,10)	132.8(5)	133.1(2)
A(5,9,11)	135.3(2)	135.9(6)
A(9,10,12)	94.4(0)	93.1(3)
A(9,10,13)	129.8(2)	129.1(1)
A(9,11,12)	87.1(6)	89.0(8)
A(10,12,11)	86.6(3)	87.1(1)
A(11,12,16)	136.5(9)	135.4(8)
A(10,13,14)	120.1(0)	118.0(4)
A(13,14,17)	106.5829	110.6(9)
D(1,2,4,5)	158.4	167.3
D(1,2,4,6)	73.0	46.6
D(2,4,5,9)	61.5	135.2
D(2,4,6,7)	155.3	102.8
D(5,4,6,7)	24.8	136.5
D(4,5,9,10)	174.6	179.3
D(5,9,10,12)	178.7	175.0
D(10,9,11,12)	0.2	0.2
D(10,9,11,15)	179.6	178.1
D(9,10,12,16)	179.8	168.8
D(15,11,12,16)	0.5	0.7
D(10,13,14,17)	179.5	104.9

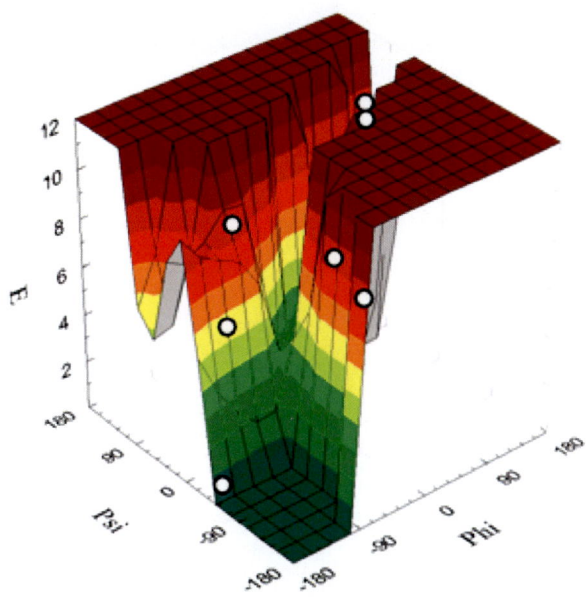

Figure 12. Landscape and contour representation of the φ and χ potential energy surphace of the molecule of *ValNHSqEt* after optimization at RHF/6-311++G**.

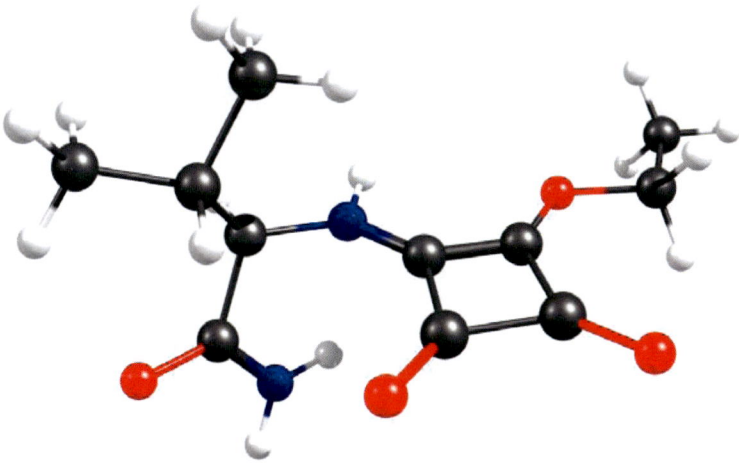

C_1 (E_{rel} = 0.2 kcal/mol).

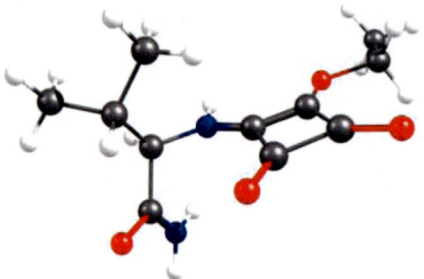

C_2 (E_{rel} = 6.1 kcal/mol).

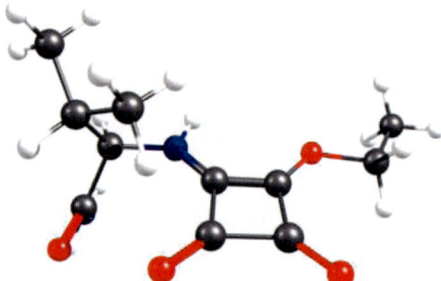

C_3 (E_{rel} = 9.3 kcal/mol).

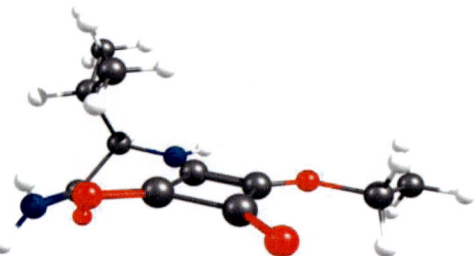

C_4 (E_{rel} = 11.7 kcal/mol).

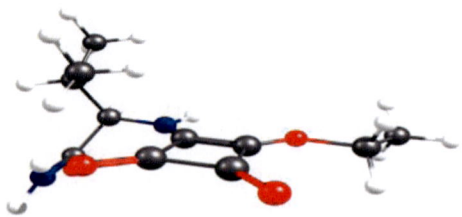

C_2 (E_{rel} = 3.2 kcal/mol).

Scheme 27. (Continued on next page.)

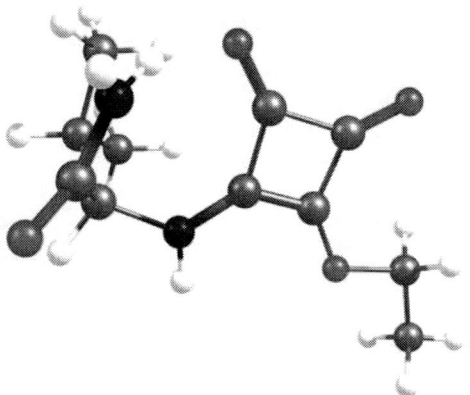

C_6 (E_{rel} = 4.1 kcal/mol).

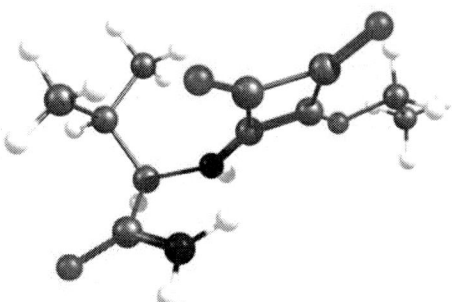

C_7 (E_{rel} = 8.8 kcal/mol).

Scheme 27. Ab initio model conformers (C_i, i = 1-7) of *ValNHSqEt* after optimization at RHF/6-311++G** level of theory.

For all he ester amides of squaric acid presented the planar amide fragment is predicted as well as the mutual co-planar disposition of NH group and the plane of squarate species. Comparing both the experimental data and theoretical here presented ones of *MetNHSqEt*, the atomic distances and angles do not differ by more than 0.099 Å and 6.3(1)°. The comparison of the experimental and theoretical predicted data indicated that the squaric acid (Sq) fragment is planar with maximal deviation within 0.9 – 2.1° and within 0.1 – 3.5° (RHF/6-31++G**). The planarity of amide fragment is also predicted with a deviation of flat configuration with 0.9o (the corresponding experimental value is 0.1°). Comparing both the experimental data and theoretical here presented ones of *PheNHSqEt* and *ValNHSqEt*, the atomic distances and angles do not differ by more than 0.031, 0.075 Å, 2.9° and 1.7(5)°, respectively. For the case of *MetNHSqEt* the theoretical data are compared with experimental ones and for other amide derivatives of methionine as N-acetyl-$_{DL}$-methionine-methyl(ethyl)amide [136] and N,N-bis(1-propan-2-onyl oxime)-$_L$-methionine N'-methylamide [137] obtaining as well a good correlations as far as the predicted geometry parameters are within 0.021-0.067 Å and 1.7(5)-2/9(8)°, respectively. In the cases of *PheNHSqEt* and *ValNHSqEt* the good correlation with experimental single crystal data for similar compounds as N-acetyl-$_{DL}$-phenylalanine-N-methylamide [138], N-acetyl-$_{DL}$-phenylalanine N',N'-dimethylamide [139], rac-Na-(t-butoxycarbonyl)-$_L$-phenylalanine N-methoxy-N-methylamide [140], aspartyl-phenylalanine amide sesquihydrate [141], Acetyl-$_L$-valine-dimethylamide [142], N(α)-acetyl-aza-α'-homo-$_L$-valine dimethylamide [143], N-((E)-1-Oxo-3-phenylprop-2-enyl)-(S)-valine amide [144] and N(α)-acetyl-aza-α'-homo-$_L$-valine methylamide [145] are obtained due to the atomic distances and angles do not differ by more than range 0.045-0.091 Å, 4.7(9)°-2.7(6)°, respectively.

Like in protonated forms of amino acid amides, independently of the used level of theory the experimental geometry in solid state is differ. In the case of *PheNHSqEt* a H$_2$O solvent molecule is included as well as the suitable proton acceptors and proton-donating groups are also obtained in the unit cell of compound studied. For this reason the stabilization of intermolecular hydrogen bond interactions are observed (Scheme 28) with types and bond lengths of: NH…O=C(NH$_2$) (2.822 Å), NH$_2$…O=C(Sq) (3.027, 2.931 Å), HOH…O=C(NH$_2$) (2.699 Å), HOH…O=C(Sq) (2.868 Å), NH…OH$_2$ (2.860), HOH…NH$_2$ (2.844 Å), respectively. These results affected on the theoretical and experimental differences in dihedral angle values vary within 7.8 – 141.9°.

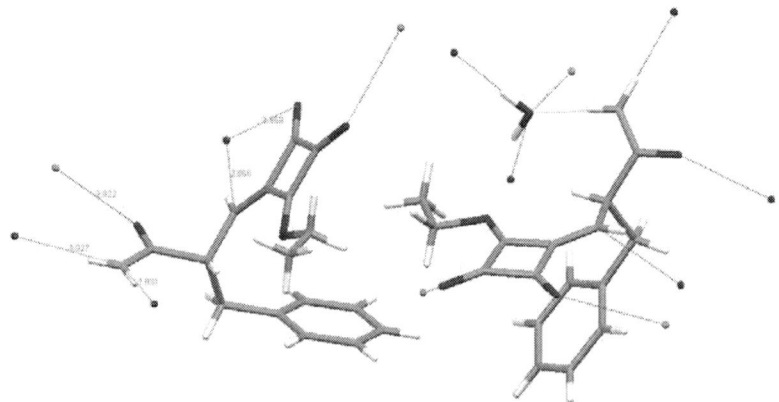

Scheme 28. Hydrogen bonds in *PheNHSqEt* [23].

In the case of *MetNHSqEt* and *ValNHSqEt* the series of intermolecular interactions are also established resulted: (Schemes 29 and 30). A significant deviation in *ValNHSqEt* is determined comparing the corresponding data about the dihedral angles in gas phase and in solid state. The calculated values of D(2,4,6,8) and D(3,2,4,5) dihedral angles (Scheme 1) of 112.1° and 177.2° are differ than experimental determined ones by single crystal X-ray diffraction equal to 75.4° and 67.4°, which has been explained by the presence of strong intermolecular interactions in solid-state leading to a deviation of theoretical predicted dihedral angles in gas phase and isolated molecule. Similar results are obtained about di- and tripeptide systems. In this case both amide and NH fragments are included in intermolecular HN-H…O, C=O…N and N-H…O interactions with bond lengths at 2.999 Å, 2.804 Å, respectively. For *MetNHSqEt* the data indicated the influence of the intermolecular interactions on the φ and χ_1 values with 4.5 and 7.9°, respectively comparing the theoretical and experimental data (tables 13 and 15).

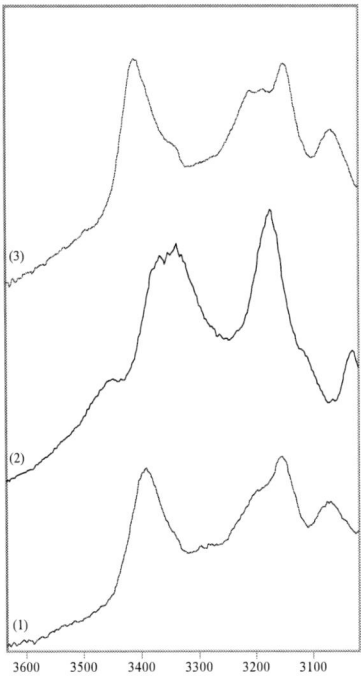

Figure 13. (Continued on next page.)

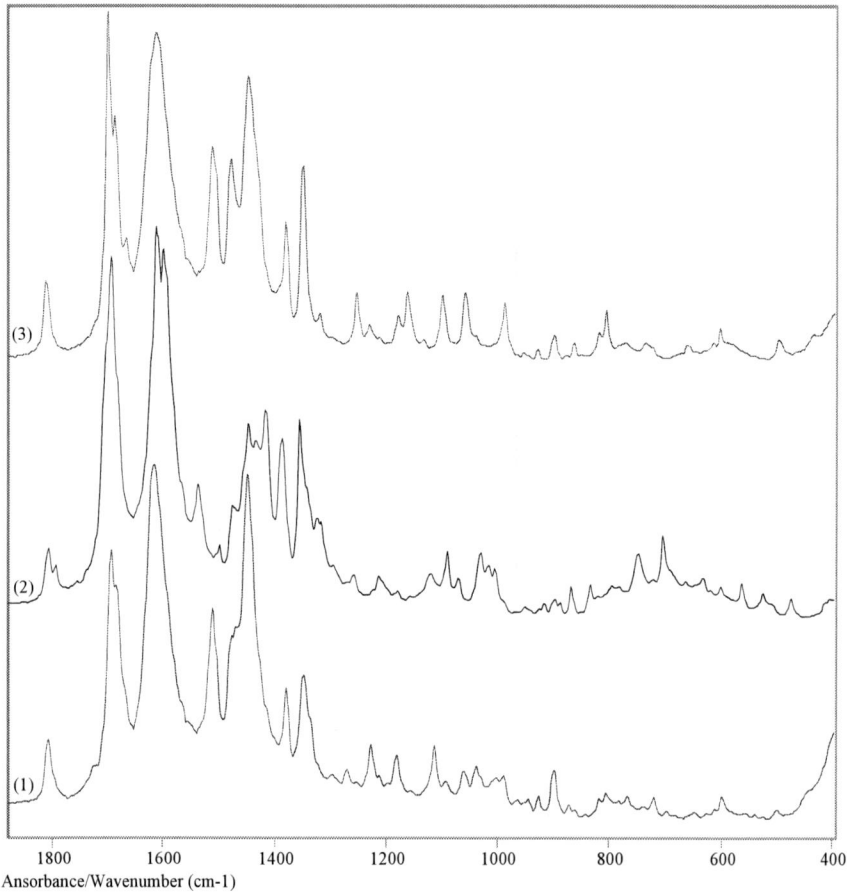

Figure 13. Solid-state IR-spectra of *MetNHSqEt* (1), *PheNHSqEt* (2) and *ValNHSqEt* (3) as KBr pellets.

The influences of intermolecular interactions on are affected as well on the IR-peak positions and integral absorbances in all the ester amides of squaric acid studied (figure 13). For *MetNHSqEt* the IR-spectroscopic characterization has been determined by linear-polarization IR-spectroscopy of solid-sample oriented by here presented method. The summarizing in table 16 data, are obtained by the following reductions:

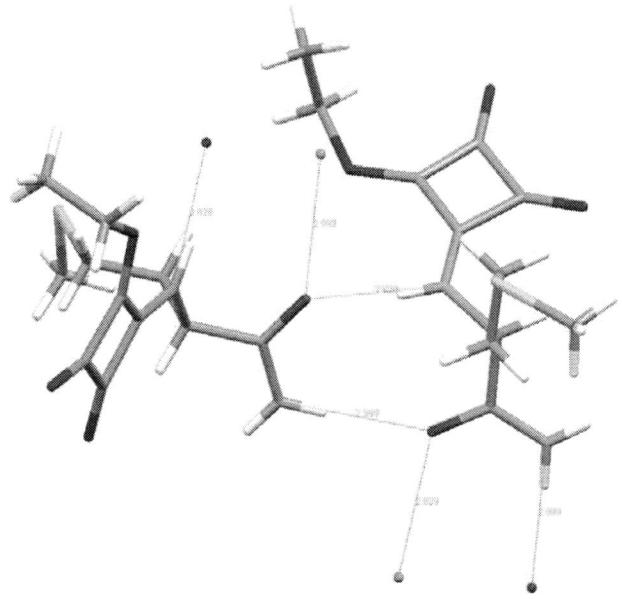

Scheme 29. Hydrogen bonds in *MetNHSqEt* [27].

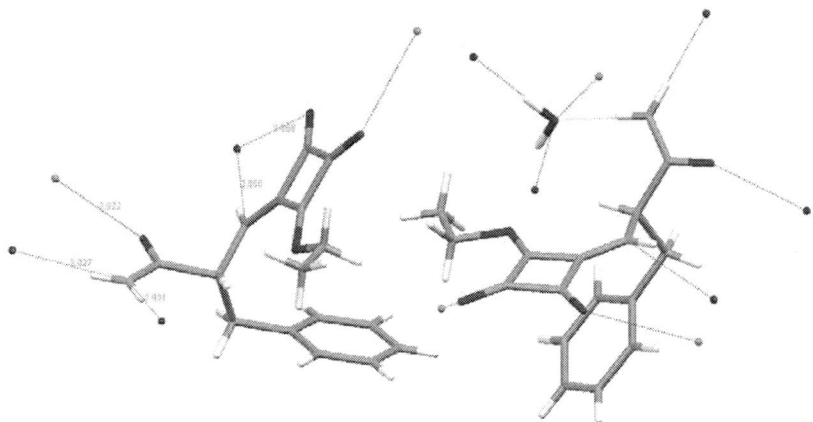

Scheme 29. Hydrogen bonds in *PheNHSqEt* [23].

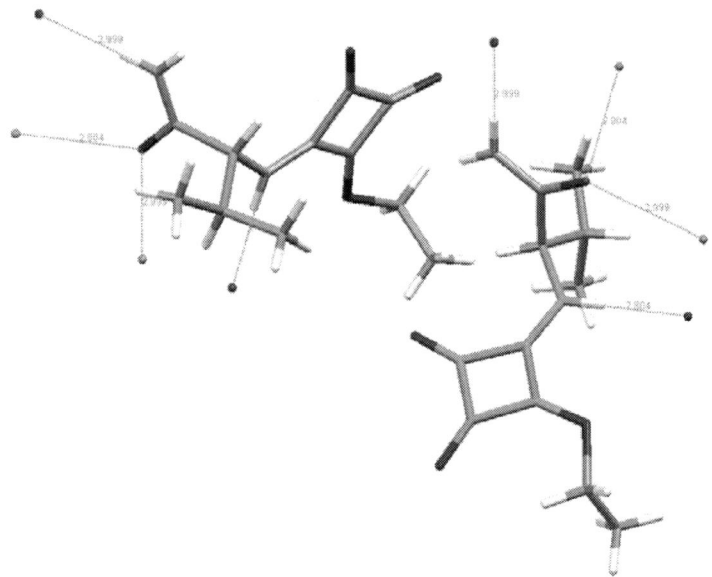

Scheme 30. Hydrogen bonds in *ValNHSqEt* [28].

Table 16. Characteristic IR-spectral bands of *MetNHSqEt* in solid state [32]

ν [cm^{-1}]	Assignment
3392	ν^{as}_{NH2}
3199	ν_{NH}
3158, 3070	ν^{s}_{NH2}
1808, 1687, 1617	$\nu^{as}_{C=O}$ (Sq), $\nu^{s}_{C=O}$ (Sq), $\nu_{C=C}$ (Sq)
1695	$\nu_{C=O}$ (Amide I)
1627	δ_{NH2}
1508	δ_{NH}
802	ρ_{CH2}(Sq)
771	ω_{NH2}
744	τ_{NH2}
597	$\gamma_{C=O}$
440	γ_{C-O}(Sq)

According [32] the elimination of the peak at 1695 cm^{-1} leads to disappearance of 3199 cm^{-1} maximum due to in the frame of one molecule *MetNHSqEt* the transition moments of Amide I ($v_{C=O}$) and v_{NH} are near to co-linear oriented (Scheme 32A, table 16). The result indicates the assignment of 3199 cm^{-1} to v_{NH} (figure 14 [32]).

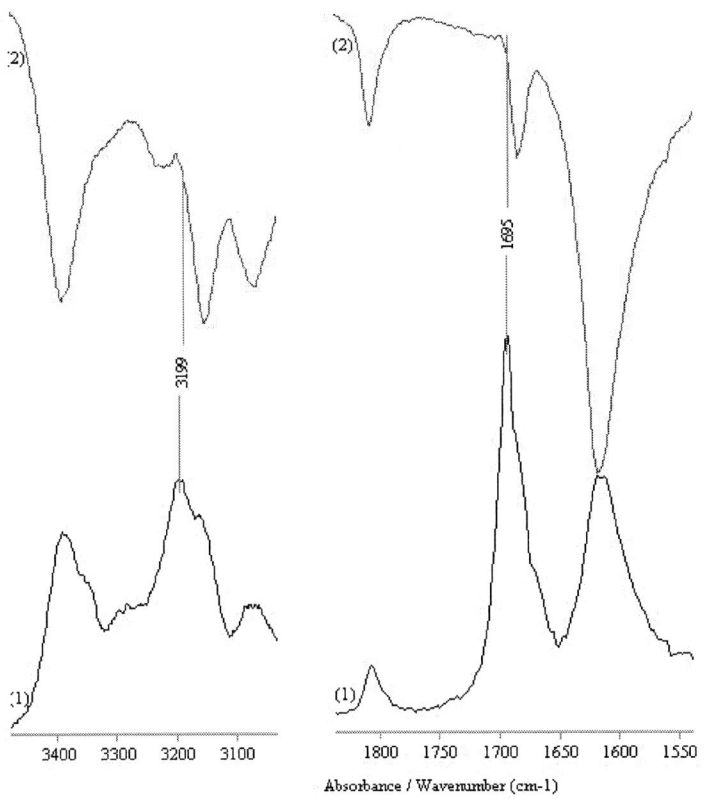

Figure 14. Non-polarized IR-(1) and reduced IR-LD spectrum of *MethNHSqEt* after elimination of the peak at 1695 cm^{-1} (2) [32].

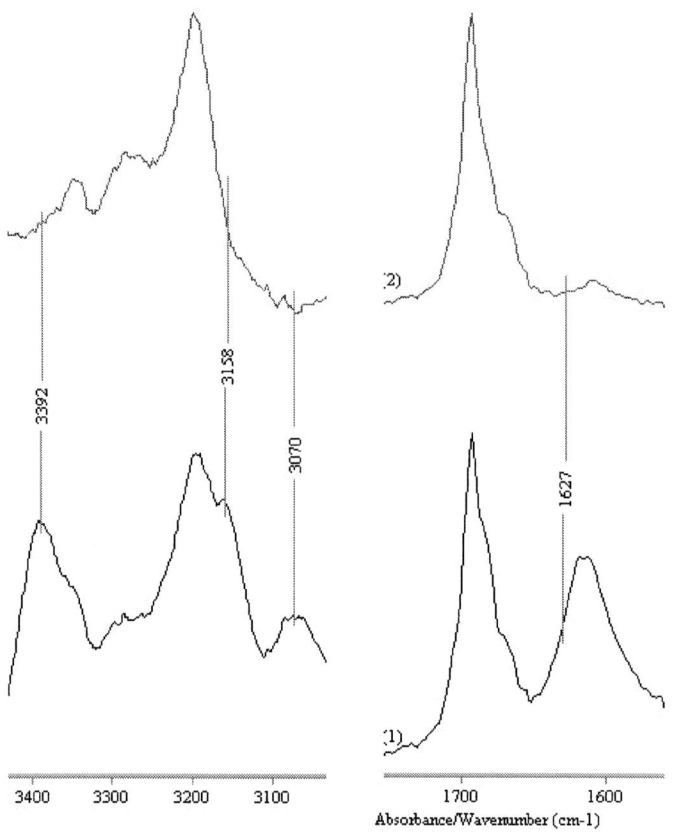

Figure 15. Non-polarized IR-(1) and reduced IR-LD spectrum of *MetNHSqEt* after elimination of the peak at 3392 cm^{-1} (2) [32].

In the frame of the unit cell of *MetNHSqEt* the two molecules are mutual oriented (Scheme 32B) supposing the near to co linearity of v^{as}_{NH2} and v^{s}_{NH2} transition moments between neighboring molecules, which explain the simultaneously disappearance of 3392 cm^{-1}, 3158 cm^{-1} and 3064 cm^{-1} peaks assigned to v^{as}_{NH2} and Fermi-resonance spitted v^{s}_{NH2} (figure 15 [32]). In parallel, a peal at 1627 cm^{-1} is also vanished clearing its δ_{NH2} character as a result of co-linear orientation of v^{s}_{NH2} and δ_{NH2} transition moments in one NH$_2$-fragment (Scheme 4A). For the purpose of the coordination ability of methionine side chains [103,105,106,146-148], the determination of the peak of v_{SC} has been achieved by the elimination of the peak of $\delta^{s}_{(S)CH3}$, which must lead to the disappearance of the discussed maximum due to their mutual co linearity (Scheme 32A). According to literature data the $\delta^{s}_{(S)CH3}$ peak is usually observed in the range of 1340 ± 40 cm^{-1} [124] and in our case the peak is at 1348 cm^{-1}. Its elimination leads to vanishing of 721 cm^{-1} peak, belonging to ν_{SC} (figure 16 [32]). The value is in accordance with the data about other methionyl-containing peptides [103,105,106] and pure amino acid [127, 149-154]. The last procedure provoked the disappearance of 1378 cm^{-1} peak, which corresponds to $\delta^s_{CH3(Sq)}$ due to the co-linearity of $\delta^s_{CH3(Sq)}$ and $\delta^s_{(S)CH3}$ transition moments in the frame of the neighboring molecules in the unit cell of *MetNHSqEt* (Scheme 32A). The elimination of the maximum at 440 cm^{-1} strong reduced the entire Sq characteristic ones in 1800 – 1450 cm^{-1} region. According to crystallographic data the Sq fragments in the two molecules of *MetNHSqEt* in the unit cell are disposed with an angle of 85.0(2)° (Scheme 32B), leading to co linearity of in-plane and out-of-plane transition moments in squarate fragments, which explained the observed elimination of $\gamma_{C-O(Sq)}$ mode (out-of-plane) and the combination in-plane ones of $\nu^{as}_{C=O\,(Sq)}$, $\nu^s_{C=O\,(Sq)}$, $\nu_{C=C\,(Sq)}$. The data correlated well with obtain for other amino-acid amides of squaric acid [23-34,111-117].

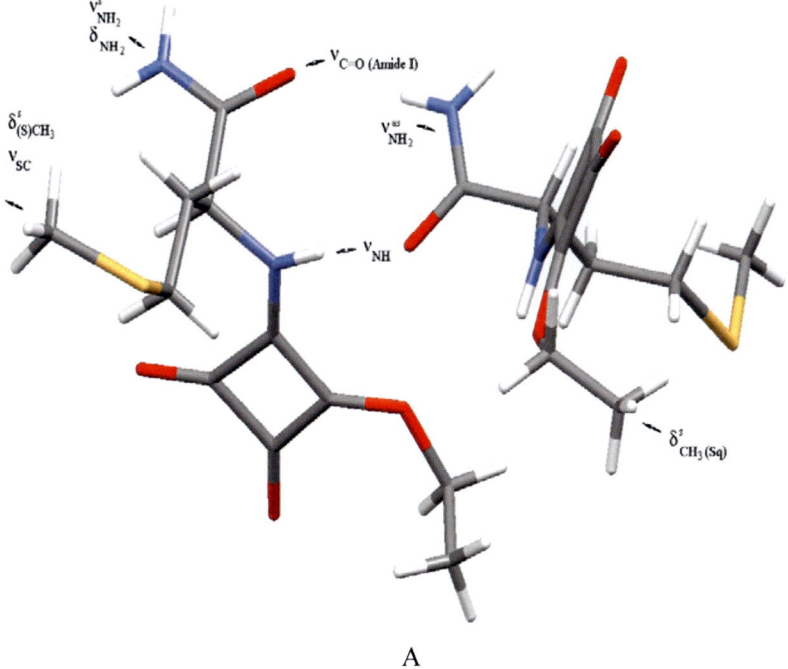

A

Scheme 32. (Continued on next page.)

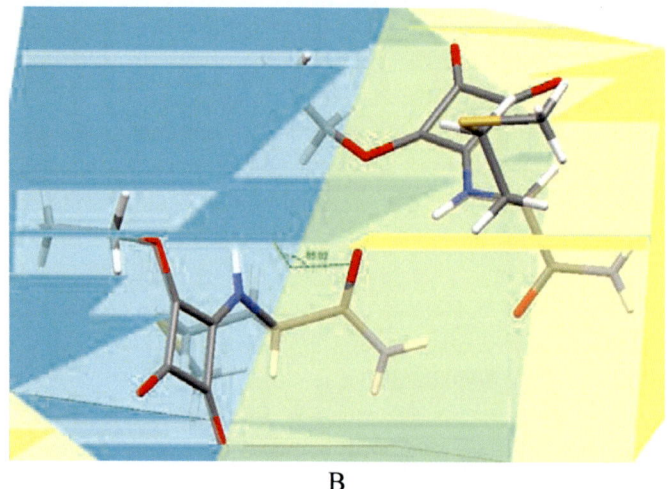

B

Scheme 32. Some transition moments (A) and mutual disposition of the molecules in the unit cell of *MetNHSqEt* [27].

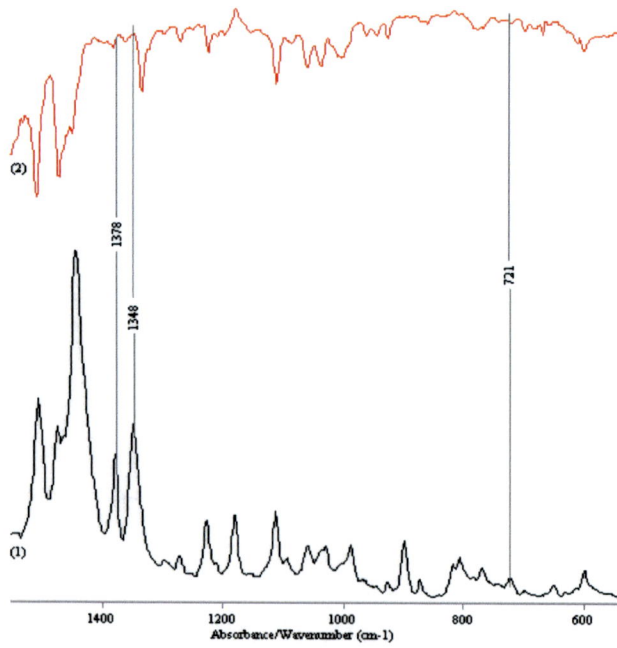

Figure 16. Non-polarized IR-(1) and reduced IR-LD spectrum of *MetNHSqEt* after elimination of the peak at 1348 cm^{-1} (2).

The simultaneously elimination of peaks at 597 cm^{-1} and 771 cm^{-1} indicated their $\gamma_{C=O}$ and ω_{NH_2} amide out-of-plane character due to the mutual co-linear transition moments in the frame of the planar amide –C=O-NH$_2$ fragment (Scheme 32B). The data correlated well with literature ones for primary amides indicating the IR-spectral ratios of 550 ± 50 cm^{-1} and 670 ± 70 cm^{-1} [124].

According to crystallographic data [23] For *PheNHSqEt* the complex IR-spectral investigation including preliminary deconvolution, curve fitting procedure and next IR-LD analysis is presented in and the IR-spectroscopic band assignment is and summarized in [135]. The characteristic IR-bands are represented in table 17 and the corresponding IR-spectrum in solid-state is illustrated in figure 13(2).

Table 17. Characteristic IR-spectral bands of *PheNHSqEt* in solid state [135]

ν [cm^{-1}]	Assignment
3517, 3463	$\nu^{as}_{H_2O}$, $\nu^{s}_{H_2O}$
3380	$\nu^{as}_{NH_2}$
3343	ν_{NH}
3177	$\nu^{s}_{NH_2}$
3033	20a* (A$_1$, i.p.)
1807/1793, 1705/1692, 1610	$\nu^{as}_{C=O\,(Sq)}$, $\nu^{s}_{C=O\,(Sq)}$, $\nu_{C=C\,(Sq)}$
1680	$\nu_{C=O}$ (Amide I)
1643	δ_{H_2O}
1621	δ_{NH_2} (Amide II)
1610	8a* (A$_1$, i.p.)
1595	19a* (A$_1$, i.p.)
1585	8b* (B$_2$, i.p.)
1536	19b* (B$_2$, i.p.)
1534	δ_{NH}
~ 750	$\tilde{\pi}\gamma_{CH}$* (B$_1$, o.p.)
~ 705	$\tilde{4}\gamma_{Ar}$* (B$_1$, o.p.)

* Wilson's notation used in ref. [49]

The curve-fitted solid-state IR-spectrum of *PheNHSqEt* (figure 17 [135]) shows pairs of maxima could be explained with the presence of non-equivalent molecules (Scheme 33) of compound studied in the unit cell of *PheNHSqEt* [23].

Similar results have been established for methionine-containing peptides [100 - 117] and proposed for $_L$-leucinamide in [35].

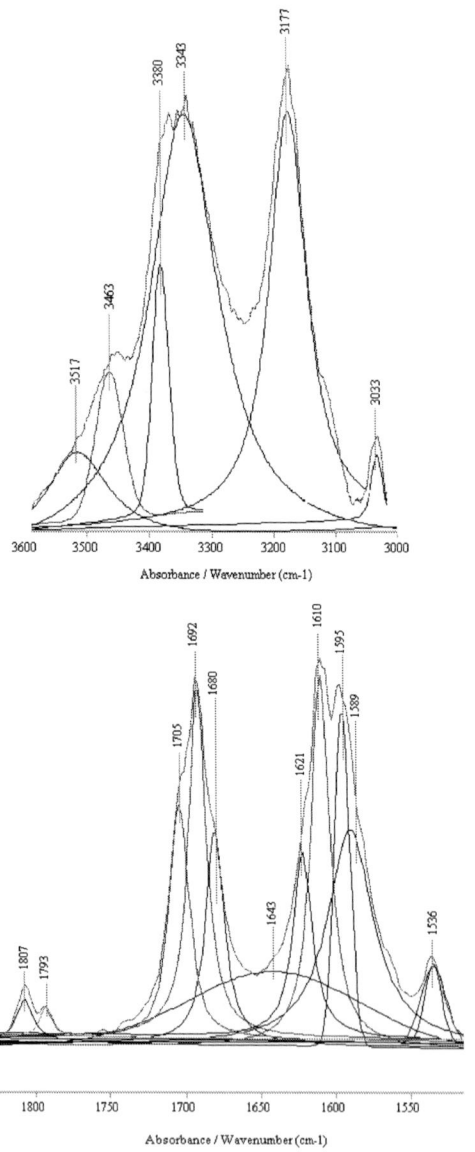

Figure 17. Curve-fitted IR-spectrum of *PheNHSqEt* [135].

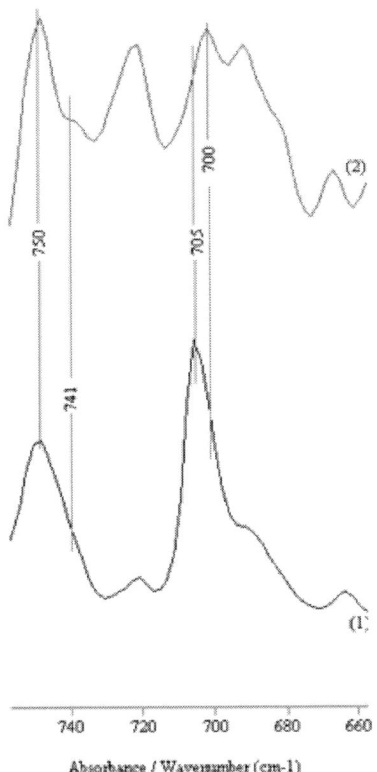

Figure 18. Non-polarized IR-(1) and reduced IR-LD spectrum of *PheNHSqEt* after elimination of the peak at 705 cm^{-1} (2).

The application of reducing difference procedure for polarized IR-spectra interpretation experimentally confirmed the IR-spectroscopic assignment stated in table 2. The elimination of 3177 cm^{-1} peak provoked the disappearance of 1621 cm^{-1}, due to the co-linear disposition of their transition moments in the frame of NH$_2$-structural fragment (Scheme 33A). On the other hand the consequently elimination of 1610 cm^{-1} and 1585 cm^{-1} resulted to a vanishing of 1595 cm^{-1} and 1536 cm^{-1} maxima due to the origin of these maxima to similar symmetry class by pairs 8a, 19a (in-plane (i.p.), A$_1$ symmetry class) and 8b, 19b (in-plane (i.p.), A$_2$ symmetry class), respectively (table 17 and Scheme 33A). In first case is eliminated as well the peak at 3033 cm^{-1}, confirming its origin as 20a i.p. mode. The elimination of $\tilde{\square}\square_{Ar}$ peak at 705 cm^{-1} leads to vanishing of 741 cm^{-1} ($11\square_{CH}$) of out-of-plane (o.p.) B$_1$ symmetry class, due to their mutual co-linear orientation (scheme 5A). In parallel the second pairs of maxima at 750 cm^{-1} and 700 cm^{-1}

corresponding to of the second molecule of *PheNHSqEt* is established (figure 18(2) [135]). This results experimentally confirmed that the observed splitting of maxima due to the non equivalent molecules in the unit cell of *PheNHSqEt* because of in this case both aromatic fragments of separate molecules of phenylalaninamide ester amide of squaric acid diethyl ester are mutual disposed at an angle of 62.3(1)° (Scheme 33B). These data correlated well with other IR- and Raman spectroscopic ones in [102-154].

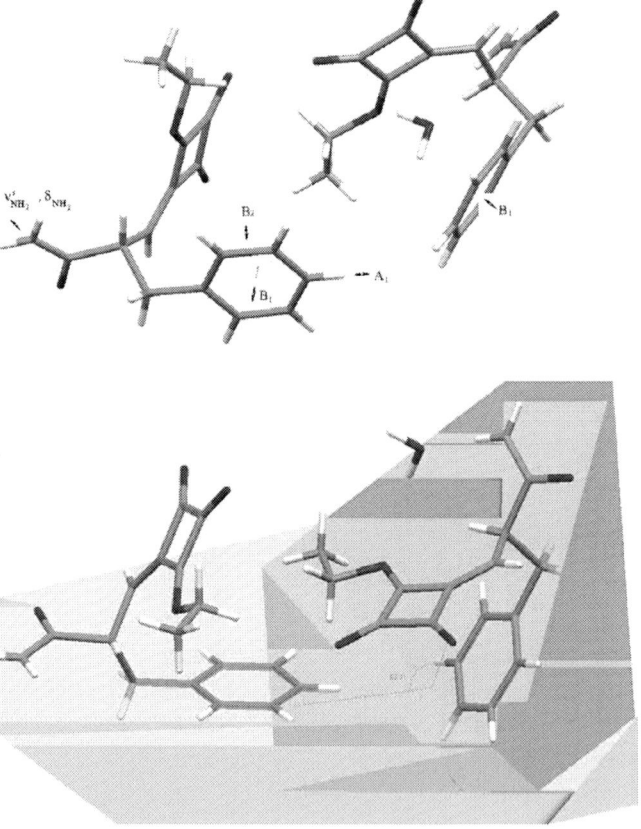

Scheme 33. Some transition moments (A) and mutual disposition of the molecules of *PheNHSqEt* in solid state [135].

In the case of *ValNHSqEt* the IR-LD data and vibrational analysis [28] could be summarized as table 18 and following reductions and assignments.

Table 18. Calculated IR-spectral frequencies of *ValNHSqEt* [28]

ν [cm^{-1}]	Assignment* (PED, %)	ν [cm^{-1}]	Assignment* (PED, %)
3514, 3477	ν^{as}_{NH2}, ν^{s}_{NH2}	1091	ρ_{NH2}
3388	ν_{NH}	1090	$\rho'_{CH3(Sq)} + \nu_{CC}$
1823, 1700	$\nu^{as}_{C=O\,(Sq)}$ and $\nu^{s}_{C=O\,(Sq)}$	1035	$\nu_{CN} + \nu_{CC}$
1690	$\nu_{C=O}$	744	τ_{NH2}
1613	$\nu_{C=C}$	719	γ/ω_{NH}
1602	δ_{NH2}	688	$\omega_{NH2} + \delta_{CO}$
1496	δ_{NH}	608	$\delta_{CO} + \omega_{NH2}$
1300	$\nu_{CN(Amide)} + \nu_{CC}$	521	$\delta_{C-O(Sq)}$
1280	$\tau_{CH2(Sq)}$	400	$\gamma_{C-O(Sq)}$

The experimental confirmation of IR-characteristic bands assignment as well as the receiving of additional structural information has been achieved by the next IR-LD spectral analysis of compound studied in solid-state, thus supporting the crystallographic data here presented. The application of reducing-difference procedure leads to following results: (i) The elimination of the peak at 1636 cm^{-1} provokes a disappearance of 3150 cm^{-1} and 3066 cm^{-1} maxima clearing their character as δ_{NH2} and Fermi-resonance spitted ν^{s}_{NH2} mode due to intermolecular interacted NH$_2$-group in solid-state, as far as in the frame of one NH$_2$-group the corresponding transition moments are co-linear oriented (Scheme 34). (ii) The reduction of $\square_{NH}$ maximum at 3200 cm^{-1} with 1810 cm^{-1} one, belonging to $\nu_{C-O(Sq)}$ and Amide I peak at 1681 cm^{-1} in same dichroic ratio, confirms the refined structure of compound studied in solid-state, as far as the corresponding co linearity of their transition moments. (iii) The vanishing of 736 cm^{-1} peak, assigned to γ_{NH} mode (theoretical value in table 18 is 719 cm^{-1}) in obtained reduced IR-LD spectrum in figure 19 [28], leads to an observation of second peak at 731 cm^{-1}, which is in accordance as well with single crystal X-ray data indicating the presence of two near to perpendicular oriented molecules in the unit cell oriented with a torsion angle of 80.9(6)°, fact supposing with the elimination of one out-of-plane mode (in this case γ_{NH}) possessing of one molecule and observation second peak with same character of second, included in the unit cell molecule of compound studied. These data are similar to IR-and Raman spectroscopic ones of valyl containing dipeptides [163-166].

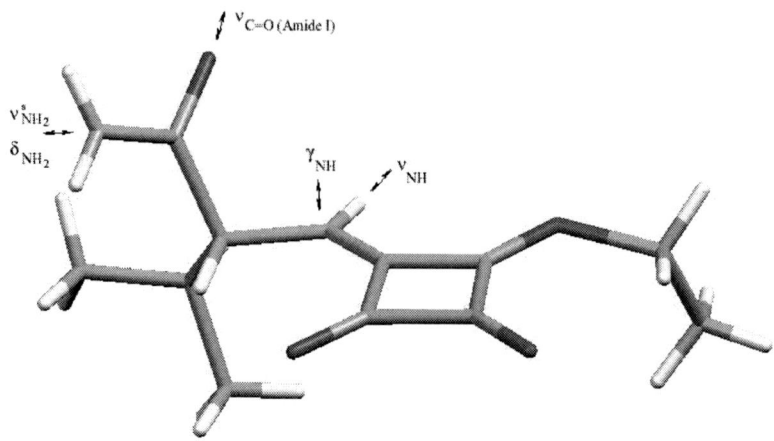

Scheme 34. Some transition moments in *ValNHSqEt* [28].

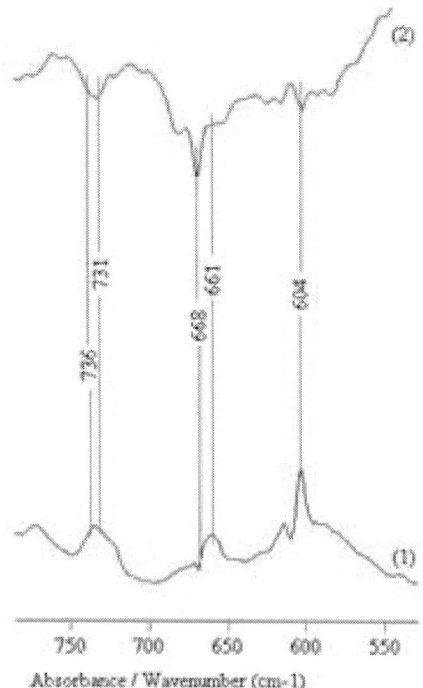

Figure 19. Non-polarized (1) and reduced IR-LD spectra of valinamide ester amide of squaric acid after elimination of peak at 604 cm^{-1} (2) [28].

Chapter 3

CONCLUSION

The presented paper illustrated a complex structural and IR-spectroscopic investigation of protonated forms of amino acid amides (argininamide, isoleucinamide, methioninamide, serinamide, tyrosinamide, tryptohanamide, threoninamide, valinamide, alaninamide and prolinamide) as hydrochlorides and hudrogensquarates as well as the ester amide derivatives of amino acid amides (methioninamide, phenylalninamide and valinamide) with squaric acid. The research is obtained by means of single crystal X-ray diffraction, linear-polarized IR-spectroscopy of oriented solid samples as suspension in nematic liquid crystal as well as the possibilities of theoretical ab initio and density functional theory calculations varying the basis sets of prediction of electronic structure and vibrational analysis.

The following mail conclusions could be made according the results obtained by above menthioned methods:

(i) The calculations at UHF and UMP2 level of theory and 6-311++G** basis set give are suitable to be used for prediction of electronic structures of protonated forms of amino acid amides as far as the comparison with experimental geometries of compound studied and similar molecules gives a bond length and angles differences do not differ within $\pm$ 3.1 - $\pm$ 1.8%. All the protonated amino acid amides stabilized in gas phase conformers with potential energy minia, characterized with presence of intramolecular hydrogen bond $N^+H_3…O=C-NH_2$ with length varying within 2.550 – 2.528 Å; the corresponding $(H_3)N^+-H-O(=C)$ angles are predicted to be in interval 112.6(9)° - 118.7(6)°, respectively. The dihedral angle values, however are significant influenced due to the formation of intermolecular interactions between the neighboring molecules of compound studied as well as with included Cl- and hydrogensquarate anions and solvent molecules. The maximal established deviation of theoretical predicted and experimental obtained dihedral angle values are within 2-25%. In all the protonated amino acid amides, the amide fragment is flat with maximal deviation of planarity varies of... The guanydil fragment in argininamide is also planar, which correlated expellant of experimental single crystal X-ray diffraction data Last fragment is characterized as well with a partial positive charge redistribution in guanidine group affected on the obtained positive singly charge values.

(ii) The good correlation between the theoretical predicted electronic structures and experimental obtained single crystal X-ray ones in amino acid amide ester amides of squaric acid using RHF and MP2 level of theory and 6-31++G** basis set. In these compound the maximal deviation between theoretical predicted and experimental obtained by single crystal X-ray data results are less than 3.8%. In all cases the most stable conformers of these compounds are characterized with planar amide fragment and co-planar disposition of NH and squarate fragments in the molecules.

(iii) DFT method provide more accurate vibrational data, as far as the calculated standard deviations 9 cm-1 (B3LYP), 22 cm^{-1} (UHF) and 21 cm^{-1}(UMP2), respectively are obtained for compound studied. So, B3LYP/6-31++G** gives a better correspondence between the experimental and theoretical values of IR-frequencies in the case when a modification of the results using the empirical scaling factor 0.9614 is applied. The obtained results correlated well with experimental ones assigned to non-affected from the hydrogen bond interaction groups in the molecules.

(iv) All the compound studied are characterized with complex experimental IR-spectroscopic patterns due to the presence of strong intermolecular hydrogen bonding between the neighboring amino acid amide molecules in the unit cell, anionic fragments and solvent molecules. The effect is significant in the cases of hydrogensquarate and ester amides of squaric acid due to the suitable proton donating and accepting atoms in squarate fragment, which could be intermolecular interacted with him. The adequate vibrational assignment of these IR-patterns is impossible for conventional IR-spectroscopic techniques. The following complex approach is appearing very suitable for spectra interpretation. The vibrational analysis at B3LYP/6-31++G** and scaling factor of 0.9614 accompanied with deconvolution and curve fitting of experimental IR-patters for determining of peak positions and next solid-state polarized IR-LD spectroscopic investigation of corresponding polarized spectra accompanied with reducing difference procedure for experimental assignment of IR-spectral bands.

The complex investigation, including quantum chemical calculations of conformational forms and the electronic structures of most stable conformer as well as vibrational analysis in gas phase; synthesis, isolation and structural characterization by single crystal X-diffraction method; spectroscopic, both conventional and linear polarized IR- and Raman characterization in solid-state of ester amides of squaric acid of prolinamide, leucinamide and tryptophanamide are in progress.

REFERENCES

[1] Lavrich, R. J., Torok, C. R., Tubergen, M. J. (2002). *J. Phys. Chem,* 106A, 8013 – 8018.
[2] Eipper, B. A., Mains, R. E. (1988). *Annu. Rev. Physiol,* 50, 333 – 344.
[3] Merkler, D. (1994). *J. Enzymme Microb. Technol,*16, 450 – 458.
[4] Suwan, S., Isobe, M., Yamashita, O., Minakata, H., Imai, K. (1994). *Insect Biochem. Mol. Biol,* 24, 1001 – 1007.
[5] Kulathila, R., Merkler, K. A., Merkler, D. J. (1999). *Nat. Prod. Rep,* 16, 145 – 151.
[6] Lavrich, R. J., Farrar, J. O., Tubergen, M. J. (1999). *J. Phys. Chem,* 103 A, 4659 – 4663.
[7] Kuhls, K. A., Centrone, C. A., Tubergen, M. J. J. (1998). *Am. Chem. Soc,* 120, 10194 –10200.
[8] In, Y., Fujii, M., Sasada, Y., Ishida, T. (2001). *Acta. Cryst,* B57, 72 – 81.
[9] Kolev, T., Spiteller, M., Sheldrick, W. S., Mayer-Figge, H., Van Almsick, T. (2006). *Acta Cryst,* E61, o3819 – o3820.
[10] Kolev, T., Spiteller, M., Sheldrick, W. S., Mayer-Figge, H. (2006). *Acta Crystallogr,* E, in preparation.
[11] Kolev, T., Spiteller, M., Sheldrick, W. S., Mayer-Figge, H. (2006). *Acta Crystallogr,* E, submitted.
[12] Wolff, J. J., Wortmann, R. (1999). *Adv. Phys. Org. Chem,* 32, 121 – 127.
[13] Chemla, D., Zyss, J. (1987). In *Nonlinear Optical Properties of Organic Molecules and Crystals*, Chemla, D., Zyss, J. Eds. Academic Press: New York, USA, Vol. 1, pp. 23 – 187.
[14] Nalwa, H. S., Watanabe, T., Miyata, S. (1997). In *Nonlinear Optics of Organic Molecules and Polymers*, Nalwa, H.S., Miyata, S. Eds. CRC Press: Boca Raton, pp. 89 – 329.

[15] Sztaricskai, F., Sum, A., Roth, E., Pelyvas, I. F., Sandor, S., Batta, G., Herczegh, P. Remenyi, J., Miklan, Z., Hudecz, F. *J. Antibiot*, 2005, 58, 704 – 714.
[16] Tevyashova, A., Sztaricskai, F., Batta, G., Herczegh, P., Jeney, A. (2004). *Bioorg. Med. Chem. Lett,* 14, 4783 – 4789.
[17] Xie, J., Comeau, A. B., Seto, C. T. (2004). *Org. Lett*, 6, 83 – 86.
[18] Kim, C. U., Misco, P. F. (1992). *Tetrahedron Lett*, 33, 3961 – 3966.
[19] Porter, J., Archibald, S., Child, K., Critchley, D., Head, J., Linsley, J., Parton, T., Robinson, M., Shock, A., Taylor, R., Warrellow, G., Alexander, R., Langham, B. (2000). *Bioorg. Med. Chem. Lett*, 12, 1051 – 1054.
[20] Onaran, M. B., Comeau, A. B., Seto C. T. (2005). *J. Org. Chem*, 70, 10792 – 10802.
[21] Gilbert, M., Antane, T., Argentier, J., Butera, G., Francisco, C., Freeden, E., Gundersen, R., Craceffa, D., Herbst, B., Hirth, J., Lennox, G., McFarlane, N. W., Norton, D., Quaglito, J., Sheldon, D., Warga, A., Wojdan, M. (2000). *J. Med. Chem,* 43, 1203 – 1214.
[22] Chan, M. C. (1995). *J. Med. Chem*, 38, 4433 – 4437.
[23] Kolev, T., Petrova, R., Spiteller, M. (2004). *Acta Cryst*, E60, o634 – o635.
[24] Kolev, T., Benet-Buchholz, J., Spiteller, M. (2006). Private communication.
[25] Kolev, T., Benet-Buchholz, J., Spiteller, M. (2006). Private communication.
[26] Kolev, T., Benet-Buchholz, J., Spiteller, M. (2006). Private communication.
[27] Kolev, T., Cherneva, E., Spiteller, M., Sheldrick, W. S., Mayer-Figge, H. (2006). *Acta Cryst*, E62, o1390 – o1391.
[28] Kolev, T., Koleva, B. B., Cherneva, E., Spiteller, M., Sheldrick, W. S., Mayer-Figge, H. (2006). *Struct. Chem*, in press.
[29] Kolev, T., Yancheva, D., Spiteller, M., Sheldrick, W. S., Mayer-Figge, H. (2006). *Acta Cryst*, E62, o463 – o464.
[30] Kolev, T., Spiteller, M., Sheldrick, W. S., Mayer-Figge, H. (2006). *Acta Cryst*, E61, o4292 – o4293.
[31] Kolev, T., Spiteller, M., Sheldrick, W. S., Mayer-Figge, H. (2006). *Acta Cryst, C62*, o299 – o300.
[32] Cherneva, E., Kolev, T. (2006). *J. Mol. Struct*, in press
[33] Kolev, T. (2006). *J. Mol. Struct*, in press.
[34] Zareva, S. (2006). *Centr. Europ. J. Chem*, in press.
[35] Wang, Y., Belton, P. S., Tang, H., Wellner, N., Davies, S. C., Hughes, D. L. *J. Chem.. Soc., Perkin Trans. 2* 1997, 899 – 904.
[36] DALTON, A molecular electronic structure program, Release 2.0, 2005 [http://www.kjemi.uio.no/software/dalton/dalton.html].

[37] Frisch, M. J., Trucks, G. W., Schlegel, H. B., Scuseria, G. E., Robb, M. A., Cheeseman, J. R., Zakrzewski, V. G., Montgomery Jr. J. A., Stratmann, R. E., Burant, J. C., Dapprich, S., Millam, J. M., Daniels, A. D., Kudin, K. N., Strain, M. C., Farkas, Ö., Tomasi, J., Barone, V., Cossi, M., Cammi, R., Mennucci, B., Pomelli, C., Adamo, C., Clifford, S., Ochterski, J., Petersson, G. A., Ayala, P. Y., Cui, Q., Morokuma, K., Salvador, P., Dannenberg, J. J., Malick, D. K., Rabuck, A. D., Raghavachari, K., Foresman, J. B., Cioslowski, J., Ortiz, J. V., Baboul, A. G., Stefanov, B. B., Liu, G., Liashenko, A., Piskorz, P., Komáromi, I., Gomperts, R., Martin, R. L., Fox, D. J., Keith, T., Al-Laham, M. A., Peng, C. Y., Nanayakkara, A., Challacombe, M., Gill, P. M. W., Johnson, B., Chen, W., Wong, M. W., Andres, J. L., Gonzalez, C., Head-Gordon, M., Replogle, E. S., Pople, J. A. (1998).Gaussian 98, Gaussian, Inc., Pittsburgh, PA.

[38] Zhurko, G. A., Zhurko, D. A. (2005).ChemCraft: Tool for treatment of chemical data, Lite version build 08 (freeware).

[39] Binkley, J. S., Pople, J. A., Hehre, W. J. (1980). *J. Am. Chem. Soc*, 102, 939 – 944.

[40] Gordon, M. S., Binkley, J. S., Pople, J. A., Pietro, W. J. Hehre, W. J. (1982). *J. Am. Chem. Soc*, 104, 2797 – 2801.

[41] Pietro, W. J., Francl, M. M., Hehre, W. J., Defrees, D. J., Pople, J. A., Binkley, J. S. (1982). *J. Am. Chem. Soc*, 104, 5039 – 5042.

[42] Dobbs K. D., Hehre, W. J. (1986). *J. Comp. Chem*, 7, 359 – 366.

[43] Dobbs K. D., Hehre, W. J. (1987). *J. Comp. Chem*, 8, 861 – 868.

[44] Dobbs K. D., Hehre, W. J. (1987). *J. Comp. Chem*, 8, 880 – 888.

[45] Ditchfield, R., Hehre, W. J., Pople, J. A. (1971). *J. Chem. Phys*, 54, 724 – 734.

[46] Hehre, W. J., Ditchfield, R., Pople, J. A. (1972). *J. Chem. Phys*, 56, 2257 - 2260.

[47] Hariharan P. C., Pople, J. A. (1974). *Mol. Phys*, 27, 209 – 211.

[48] Gordon, M. S. (1980). *Chem. Phys. Lett*, 76, 163 – 166.

[49] Hariharan, P. C., and Pople, J. A. (1973). *Theo. Chim. Acta*,28, 213 – 219.

[50] Blaudeau, J. P., McGrath, M. P., Curtiss, L. A., Radom, L. (1997). *J. Chem. Phys*, 107, 5016 – 5019.

[51] Francl, M. M., Pietro, W. J., Hehre, W. J., Binkley, J. S., DeFrees, D. J., Pople, J. A., Gordon, M. S. (1982). *J. Chem. Phys*, 77, 3654 – 3662.

[52] Binning Jr. R. C., Curtiss, L. A. (1990). *J. Comp. Chem,*11, 1206 – 1210.

[53] Rassolov, V. A., Pople, J. A., Ratner, M. A., Windus, T. L. (1998). *J. Chem. Phys*, 109, 1223 – 1227.

[54] Rassolov, V. A., Ratner, M. A., Pople, J. A., Redfern, P. C., Curtiss, L. A. (2001). *J. Comp. Chem*, 22, 976 – 982.
[55] Petersson, G. A., Bennett, A., Tensfeldt, T. G., Al-Laham, M. A., Shirley, W. A., Mantzaris, J. (1988). *J. Chem. Phys*, 89, 2193 – 2197.
[56] McLean A. D., Chandler, G. S. (1980). *J. Chem. Phys*, 72, 5639 – 5641.
[57] Krishnan, R., Binkley, J. S., Seeger, R., Pople, J. A. (1980). *J. Chem. Phys*, 72, 650 – 654.
[58] Wachters, A. J. H. (1970). *J. Chem. Phys*, 52, 1033 - 1037.
[59] Hay, P. J. (1977). *J. Chem. Phys*, 66, 4377 – 4381.
[60] Raghavachari, K., Trucks, G. W. (1989). *J. Chem. Phys*, 91, 1062 – 1067.
[61] Curtiss, L. A., McGrath, M. P., Blaudeau, J. P., Davis, N. E., Binning Jr. R. C., Radom, L. (1995). *J. Chem. Phys*, 103, 6104 - 6107.
[62] McGrath, M. P., Radom, L. (1991). *J. Chem. Phys*, 94, 511 – 519.
[63] Backe, D. (1993). *J. Chem. Phys*, 98, 5648 – 5656.
[64] Lee, C., Yang, W., Parr, R.G. (1988). *Phys. Rev*, B37, 785 – 803.
[65] Peng, C., Ayala, P. Y., Schlegel, H. B., Frisch, M. J. (1996). *J. Comp. Chem,* 17, 49 – 53.
[66] Head-Gordon, T., Head-Gordon, M., Frisch, M. J., Brooks III, Ch. L., Pople, J. A. (1991). *J. Am. Chem. Soc*, 113, 5989 – 5997.
[67] Vargas, R., Garza, J., Hay, B. P., Dixon, D. A. (2002). *J. Phys. Chem*, 106A, 3213 – 3218.
[68] Ramachandran, G. N., Sasisekharan, V. (1968). *Adv. Protein Chem*, 23, 283 – 438.
[69] Zimmerman, S. S., Pottle, M. S., Nemethy, G., Scherana, H. A. (1977). *Macromolec*, 10, 1 – 9.
[70] Herzberg, O., Moult, J. (1991). *Proteins*, 11, 223 – 229.
[71] Tarakeshwar, P., Manogaran, S. (1996). *J. Mol. Struct*, 365, 167 – 181.
[72] Sapse, A. M., Mallah-Levy, L., Daniels, S. B., Erickson, B. W. (1987). *J. Am. Chem. Soc*, 109, 3526 – 3529.
[73] Cioslowski, J. (1989). *J. Am. Chem. Soc*, 111, 8333 – 8339.
[74] Kolev, T. (2006). *Biopolymerts,* in press.
[75] Kolev, T., Benet-Buchholz, J., Spiteller, M. (2006). *Private communication.*
[76] Kolev, T., Benet-Buchholz, J., Spiteller, M. (2006). *Private communication.*
[77] Burns, C. S., Aronoff-Spencer, E., Dunham, C. M., Lario, P., Avdievich, N. I., Antholine, W. E., Olmstead, M. M., Vrielink, A., Gerfen, G. J., Peisach, J., Scott, W. G., Millhauser G. L. (2002). *Biochemistry,* 41, 3991 – 4000.
[78] Harada, Y., Iitaka Y. (1977). *Acta Cryst*, 33B, 244 – 253.
[79] Karle I. L. (1980). *ACA, Ser,* 2, 7, 25 – 28.

References

[80] Souhassou, M., Aubry, A., Lecomte C. (1990). *Acta Cryst*, 46C, 1303 – 1305.
[81] Simpson, P. G., Kahrl J. H. (1973). *ACS, Abstr.Papers (Summer)*, 63.
[82] Coll, M., Subirana, J. A., Solans, X., Font-Altaba, M., Mayer R. (1987). *Int. J. Pept. Protein Res*, 29, 708 – 711.
[83] Ivanova, B. B., Arnaudov, M. G., Bontchev, P. R. *Spectrochim. Acta*, 2004, 60A, 855 – 862.
[84] Arnaudov, M. G., Ivanova, B. B., Dinkov, Sh. G. (2005). *Vibr. Spectrosc*, 37, 145 – 147.
[85] Ivanova, B.B., Mayer-Figge, H. (2005). *J. Coord. Chem*, 58, 653 – 659.
[86] Koleva, B. B., Trendafilova, E. N., Arnaudov, M. G., Sheldrick, W. S., Mayer-Figge, H. (2006). *Trans. Met. Chem*, in press.
[87] Ivanova, B. B., Arnaudov, M. G., Mayer-Figge, H. (2005). *Polyhedron*, 24, 1624 – 1630.
[88] Ivanova, B. B. (2005). *Spectrochim. Acta*, 62A, 58-65.
[89] Ivanova, B. B. (2005). *Spectrosc. Lett*, 38, 635 – 643.
[90] Arnaudov, M. G., Ivanova, B. B. (2005). *Bulg, Chem. Commun*, 37, 283 – 288.
[91] Ivanova, B. B., Pindeva, L. I. (2006). *J. Mol. Struct*, in press.
[92] Ivanova, B. B. (2005). *J. Mol. Struct*, 738, 233 – 238.
[93] Koleva, B. B. (2006). *J. Mol. Struct*, in press.
[94] Ivanova, B. B. (2006). *Centrl. Europ. J. Chem*, 4, 111 – 117.
[95] Ivanova, B. B. (2006). *Vibr. Spectr*, submitted.
[96] Ivanova, B. B. (2006). *Spectrochim. Acta*, Part A, in press.
[97] Kolev, T., Ivanova, B. B., Bakalska, R. (2006). *J. Mol. Struct*, in press.
[98] Ivanova, B. B., Kolev, T., Bakalska, R. (2006). *Spectrochim. Acta*, Part A, in press.
[99] Kolev, T., Koleva, B. B., Emgenbroich, M., Spiteller, M., Sheldrick, W. S., Mayer-Figge, H. (2006). *Struct. Chem*, submitted.
[100] Ivanova, B. B. (2005). *J. Coord. Chem*. 58, 587 – 593.
[101] Arnaudov, M. G., Ivanova, B. B., Todorov, St. T., Zareva, S. Io. (2006). *Spectrochim. Acta* 63A, 491 – 500.
[102] Ivanova, B. B., Arnaudov, M. G. (2006). *Prot. Pept. Lett*, in press.
[103] Ivanova, B. B., Arnaudov, M. G., Todorov, St. T. (2006). *J. Coord. Chem*, in press.
[104] Ivanova, B. B. (2006). *J. Mol. Struct*, 782, 122 – 129.
[105] Ivanova, B. B., Arnaudov, M. G. (2006). *Spectrochim. Acta*, Part A, in press.

[106] Ivanova, B. B., Arnaudov, M. G., Todorov, S.T., Sheldrick, W. S., Mayer-Figge, H. (2006). *Struct. Chem*, 17, 49 – 56.
[107] Ivanova, B. B. (2006). *Spectrochim. Acta*, 64A, 931 – 938.
[108] Ivanova, B. B. (2006). *Spectrochim. Acta*, Part A, submitted.
[109] Ivanova, B. B., Kolev, T., Zareva, S. Y. (2006). *Biopolymers*, in press.
[110] Kolev, T., Ivanova, B. B., Zareva, S. Y. (2006). *J. Coord. Chem*, submitted.
[111] Koleva, B. B., Kolev, T.S., Zareva, S. Y., Spiteller, M. (2006). *J. Mol. Struct*, submitted.
[112] Koleva, B. B., Kolev, T., Zareva, S. Y., Spiteller, M. (2006). *J. Mol. Struct*, submitted.
[113] Kolev, T., Koleva, B. B., Spiteller, M. (2006). *Amino Acids*, submitted.
[114] Koleva, B. B. (2006). *Vibr. Spectrosc*, in press.
[115] Kolev, T.S., Zareva, S. Y., Koleva, B. B., Spiteller, M. (2006). *Inorg. Chim. Acta*, in press.
[116] Kolev, T. M., Koleva, B. B., Spiteller, M. (2006). *Biopolymers*, submitted.
[117] Koleva, B. B., Kolev, T. M., Spiteller, M. (2006). *Biopolymers*, submitted
[118] Ivanova, B. B., Tsalev, D. L., Arnaudov, M. G. (2006). *Talanta*,69, 822 – 828.
[119] Ivanova, B. B., Simeonov, V. D., Arnaudov, M. G., Tsalev, D. L. (2006). *Spectrochim Acta*, Part A, in press.
[120] Thulstrup, E. W. (1986). *Spectroscopy with Polarized Light. Solute alignment by photoselection, in liquid crystals, polymers and membranes*; VCH Publishers: New York,USA, pp 32 – 123.
[121] Thulstrup, E.W., Eggers, J.H. (1966). *Chem. Phys. Lett*, 1, 690 – 695.
[122] Jordanov, B., Schrader, B. (1995). *J. Mol. Struct*, 347, 389 – 395.
[123] Jordanov, B., Nentchovska, R., Schrader, B. (1993). *J. Mol. Struct*, 297, 401 – 406.
[124] Roeges, N. P. (1993). *A Guide to the Complete Interpretation of Infrared Spectra of Organic Structures*; Wiley: New York, USA, pp 1 – 256.
[125] RLima, J. C., Freire, P. T. C., Sasaki, J. M., Melo F. E. A., Filho, J. M. (2002). *J. Raman Spectrosc*, 33, 625 – 630.
[126] Ramaswamy, S., Rajaram, R. K., Ramakrishnan, V. (2003). *J. Raman Spectrosc*, 34, 50 – 56.
[127] Cao, X., Fischer, G. (2002). *J. Phys. Chem, A*, 106, 41 – 50.
[128] Onoa, B., Moreno, V. (1998). *Trans. Met. Chem*, 23, 485 – 490.
[129] Tandon, P., Gupta, V. D., Prasad, O., Rastogi, S., Gupta, V. D. (1997). *J. Polym. Sci. B: Polym. Phys*, 35, 2281 – 2288.
[130] Gatlin, C. L., Turecek, F., Vaisar, T. (1995). *J. Mass Spectrom*, 30, 1617 – 1619.

References

[131] Grochowski, T., Samochocka, K. (1992). *J. Chem. Soc., Dalton Trans*, 1145 – 1149.
[132] Isab, A. A. (1988). *Inorg. Chim. Acta,*153, 209 – 215.
[133] Kowalik, T., Kozlowski, H. (1982). *Inorg. Chim. Acta,*67, L39 – L41.
[134] Scott, A. P., Radom, L. (1996). *J. Phis. Chem*, 100, 16502 – 16510.
[135] Kolev. T.S. (2006). *Amino acids,* submitted.
[136] Aubry, A., Protas, J., Cung, M. T., Marraud M. (1979). *Acta Cryst,* 35B, 2634 – 2639.
[137] Goldcamp, M. J., Rosa, D. T., Landers, N. A., Mandel, S. M., Krause Bauer J. A., Baldwin M. J. (2000). *Synthesis,*2033 – 2041.
[138] Harada, Y., Iitaka Y. (1974). *Acta Cryst,* 30, 726 – 730.
[139] Siodlak, D., Broda, M. A., Rzeszotarska, B., Dybala, I., Koziol A. E. (2003). *J. Pept. Sci,* 9, 64 – 71.
[140] Zheng, X., Donkor, I. O., Miller, D. D., Ross C. R. (2000). *Chirality,* 12, 2 – 12.
[141] In, Y., Tani, S., Ishida T. (2000). *Chem. Pharm. Bull,* 48, 374 – 380.
[142] Aubry, A., Cung, M. T., Marraud M. (1982). *Cryst. Struct. Commun,*11,129 – 130.
[143] Mishnev, A. F., Bleidelis, Ya. Ya., Antsans, Yu. E., Chipens G. I. (1982). *Zh. Strukt. Khim. (Russ.) (J.Struct.Chem.),*23, 101 – 102.
[144] Baures, P. W., Beatty, A. M., Dhanasekaran, M., Helfrich, B. A., Perez-Segarra, W., Desper J. (2002). *J. Am. Chem. Soc,*124, 11315 – 11319.
[145] Kemme, A. A., Shvets, A. E., Bleidelis, Ya. Ya., Antsans, Yu. E., Chipens G. I. (1976). *Zh. Strukt. Khim. (Russ.) (J.Struct.Chem.),*17, 1132 – 1139.
[146] Barnham, K. J., Djuran, M. I., Murdoch, P. D. S., Ranford, J. D., Sadler, P. J. (1995). *J. Chem. Soc., Dalton Trans,* 3721 – 3726.
[147] Dalhus, B., Go¨rbitz, C. H. (1996). *Acta Chem. Scand,* 50, 544 – 549.
[148] Torii, K., Iitaka, Y. (1973). *Acta Cryst,* B29, 2799 – 2805.
[149] Onoa, B., Moreno, V. (1998). *Trans. Met. Chem,* 23, 485 – 490.
[150] Tandon, P., Gupta, V., Prasad, O., Rastogi, S., Gupta, V. (1997). *J. Polym. Sci. B: Polym. Phys,* 35, 2281 – 2288.
[151] Gatlin, C. L., Turecek, F., Vaisar, T. (1995). *J. Mass Spectrom,* 30, 1617 – 1622
[152] Grochowski, T., Samochocka, K. L. (1992). *J. Chem. Soc., Dalton Trans,* 1145 – 1150.
[153] Isab, A. (1988). *Inorg. Chim. Acta,* 153, 209 – 219.
[154] Kowalik, T., Kozlowski, H. (1982). *Inorg. Chim. Acta,* 67, L39 – L43.
[155] Przybylski, P., Brzezinski, B. (2003). *J. Mol. Struc,* 654, 167–176

[156] Rajkumar, B. J. M., Ramakrishnan, V. (2002). *Spectrochimica Acta,*58A, 1923 – 1934.
[157] Fodor, S. P. A., Copeland, R. A., Gryon, C. A., Spiro, T. G. (1989). *J. Am. Chem. Soc*, 111, 5509 – 5513.
[158] Naik, V. M. (1993). *Int. J. Pept. Protein Res*, 42, 2125 – 2150.
[159] Pessoa, J. C., Cavaco, I., Correia, I., Duarte, M. T., Gillard, R. D., Henriques, R.T., Higes, F. J., Madeira, C., Tomaz, T. (1999). *Inorg. Chim. Acta*, 293, 1 – 7.
[160] Ota, F., Higuchi, S., Gohshi, Y., Furuya, K., Ban, M., Kyoto, M. (1997). *J. Raman Spectrosc*, 28, 849 – 852.
[161] Murakami, T., Orihashi, Z., Kikuchi, Y., Igarashi, S., Yukawa, Y. (2000). *Inorg. Chim. Acta,*303, 148 – 155.
[162] Gerhards, M.; Unterberg, C. (2002). *Phys. Chem. Chem. Phys*, 4, 1760 – 1765.
[163] Jacob, R., Fischer, G. (2003). *J. Phys. Chem,* A,107, 6136 – 6143.
[164] Jacob, R., Fischer, G. (2002). *J. Mol. Struct*, 613, 175 – 182.
[165] Fischer, G., Jacob, R., Cao, X. (2001). *Chem. Phys*, 263, 243 – 247.
[166] Rajkumary, B. J. M., Ramakrishnan, V. (2000). *J. Raman Spectrosc*, 31, 1107–1112.

INDEX

A

acceptors, 64
accuracy, 48
acid, vii, 1, 2, 4, 7, 8, 16, 28, 63, 66, 71, 76, 78, 79, 80, 81
acidity, vii, 2
activation, 2
adriamycin, 2
agent, vii, 2
alanine, 4, 11
algorithm, 9
alternative, 2
amide, vii, 1, 7, 8, 10, 15, 28, 31, 33, 43, 47, 48, 63, 65, 73, 76, 78, 79, 80
amine, 11
amino, vii, 1, 2, 4, 8, 10, 11, 15, 16, 28, 31, 33, 37, 43, 64, 71, 79, 80
amino acid, vii, 1, 2, 4, 8, 10, 15, 16, 28, 31, 33, 37, 43, 64, 71, 79, 80
amino acid side chains, 28
amino acids, vii, 1, 2, 4, 10, 37
amino groups, 15
amorphous, 37
anions, 80
antagonist, vii, 2
antibiotics, 2
antitumor agent, 2
application, vii, 2, 75, 77
arginine, 4, 10, 15, 34, 43
aromatic, 29, 34, 52, 76

aspirin, 37
assignment, viii, 37, 40, 43, 69, 73, 75, 77, 81
asthma, 3
atomic distances, 63
atoms, 28, 81

B

bacterial, 3
basis set, viii, 9, 15, 29, 32, 34, 48, 79, 80
bending, 47
Bessel, 42
binding, 1, 28
binding energies, 1
bioactive, 2
biochemical, 1
biological, vii, 1, 3, 40
biological activity, vii
biologically, vii, 1
biomolecules, 1
biopolymers, 88
biosynthesis, 3
bonding, 1, 21, 28, 36, 80
bonds, 1, 32, 33, 64, 67, 68
building blocks, 1

C

calcitonin, vii, 1
cancer, 2
carboxylic, vii, 2

cell, vii, 1, 3, 16, 27, 28, 29, 31, 33, 64, 70, 72, 73, 76, 77, 80
chemical, viii, 1, 85
chemical properties, viii
chemistry, 2
chemotherapy, 2
chloride, 29
cleavage, 2
C-N, 11, 15, 31, 33, 35, 43, 79
communication, 84, 86
complexity, 43
composition, 3
compounds, viii, 3, 9, 28, 37, 64, 80
configuration, 63
conformational, viii, 2, 3, 4, 8, 9, 11, 48, 59, 81
conformational analysis, viii, 9, 11, 48
connectivity, 11
coordination, 70
copper, 29
correlation, 9, 43, 64, 80
correlation function, 9
correlations, 64
crystal, vii, 3, 8, 16, 28, 29, 34, 37, 38, 39, 40, 47, 48, 52, 64, 65, 77, 79, 80, 81
crystal structure, 28
crystalline, 3, 37
crystallographic, 32, 71, 73, 77
curve-fitting, 39, 40

D

deconvolution, 39, 40, 45, 73, 81
definition, 22, 23, 24, 25, 26, 27, 34, 58, 59, 60
degree, viii, 40
density, 9, 79
density function theory (DFT), viii, 3, 9, 32, 48, 79, 80
derivatives, vii, 1, 2, 8, 11, 29, 37, 64, 79
deviation, 15, 16, 27, 28, 29, 31, 32, 34, 47, 63, 65, 80
diabetes, 2
diffraction, vii, 16, 81

dihedral angles, 16, 22, 23, 24, 26, 29, 31, 34, 65
dipeptides, 77
diseases, 2
disposition, 10, 39, 63, 72, 75, 76, 80
DNA, vii, 2
DNA polymerase, vii, 2
drugs, 2

E

eigenvalues, 9
electronic, viii, 28, 34, 37, 79, 80, 81, 84
electronic structure, viii, 28, 34, 37, 79, 80, 81, 84
electro-optical properties, vii, 2
energy, 10, 11, 15
envelope, 11
enzymes, 2
ester, vii, 3, 4, 7, 8, 63, 66, 76, 78, 79, 80, 81
evidence, 33
experimental condition, 37, 39
experimental design, 39

F

Fermi, 40, 70, 77
flexibility, 11
FTIR, 3

G

gas, 2, 3, 8, 9, 15, 27, 34, 52, 65, 79, 81
gas phase, 2, 3, 8, 9, 15, 27, 34, 52, 65, 79, 81
gastrin, vii, 1
Gaussian, 9, 39, 85
geometrical parameters, 9
glutamate, vii, 2
glycine, 2, 11, 37
glycoside, 2
glycyl, 29
grouping, 28
groups, 27, 28, 29, 32, 33, 43, 48, 64, 80

Index

H

H_2, 11, 15, 31, 33, 35, 45, 47, 48, 79
hormones, vii, 1
human, 2
hydrogen, viii, 1, 8, 10, 11, 21, 27, 28, 29, 30, 31, 32, 33, 35, 36, 43, 47, 52, 64, 79, 80
hydrogen bonds, viii, 1, 8, 11, 27, 28, 29, 30, 31, 32, 33, 34, 35, 36, 43

I

identification, 1
in vitro, vii, 2
in vivo, 1, 2
infection, 2
infrared, 48
inhibitors, vii, 2
insight, 1
interaction, 11, 28, 33, 43, 47, 80
interactions, 1, 11, 16, 28, 29, 31, 33, 35, 40, 43, 48, 64, 65
intermolecular, viii, 1, 3, 8, 16, 27, 28, 29, 31, 33, 35, 37, 40, 43, 47, 64, 65, 66, 77, 80
intermolecular interactions, 3, 16, 28, 29, 31, 33, 35, 37, 47, 65, 66, 80
interpretation, viii, 37, 39, 75, 81
interval, 80
ionotropic glutamate receptor, vii, 2
ions, 28
IR, viii, 3, 8, 37, 38, 39, 40, 41, 42, 43, 44, 45, 46, 47, 66, 68, 69, 70, 72, 73, 74, 75, 76, 77, 78, 79, 80, 81
IR-spectra, 3, 37, 39, 40, 41, 43, 44, 45, 46, 66, 68, 73, 75, 77, 81
IR-spectroscopy, 3, 39, 66, 79
isolation, 81
isoleucine, 4

K

KBr, 37, 39, 44, 66
kinks, 11

L

lead, 28, 70
linear, viii, 3, 37, 39, 47, 66, 69, 70, 73, 75, 77, 79, 81
linkage, vii, 2
liquid crystals, 88
literature, 70, 73

M

matrix, vii, 3, 10
medication, 3
medications, vii, 2
medicinal, 2
membranes, 88
Merck, 37
methionine, 4, 64, 70, 74
methylene, 11
methylene group, 11
model system, 3, 29
models, 1, 10
moieties, 28
molecules, 1, 2, 3, 15, 27, 28, 29, 31, 33, 37, 70, 72, 73, 76, 77, 79, 80
multiple sclerosis, 3

N

N-acety, 29, 34, 64
nematic, 3, 8, 37, 79
nematic liquid crystal, 3, 8, 37, 79
nematic liquid crystals, 37
neuropeptides, vii, 1
New York, 83, 88
nitrogen, 2, 11
NMDA, 2
NMR, viii, 3
nonlinear, vii, 2
normal, viii, 3, 8

O

optical, vii, 2
optimization, 9, 12, 14, 39, 49, 52, 54, 57, 61, 63
organic, 15, 37
organic compounds, 15
orientation, 37, 39, 70, 75
orthorhombic, 31, 33
oxidative, 2
oxygen, 11

P

paper, 4, 79
parameter, 9, 39
penicillin, 3
peptide, vii, 1, 2, 10
peptide chain, 2
peptides, vii, 1, 2, 11, 37, 71, 74
perchlorate, 34
perturbation, 9
perturbation theory, 9
phenylalanine, 4, 64
phosphate, vii, 2
physiology, 3
planar, 15, 34, 63, 73, 80
play, 2, 11
polarization, 66
polarized, viii, 3, 37, 39, 40, 47, 69, 70, 72, 75, 78, 79, 81
polarized light, 39
polyamine, vii, 2
polymers, 88
polypeptides, 1, 3
potassium, 2
potential energy, 9, 11, 49, 52, 54, 59, 61, 79
power, 43
prediction, 9, 15, 79
preference, 8
preparation, 83
procedures, 37, 39, 45
program, 9, 39, 84
protein, vii, 2, 11
protein tyrosine phosphatases, vii, 2
proteins, 1, 2, 86
puckering, 11

Q

quantum, 1, 9, 48, 81
quantum chemical calculations, 9, 48, 81

R

Raman, viii, 3, 43, 76, 77, 81, 88, 90
Raman spectroscopy, viii
range, 9, 15, 37, 64, 71
receptors, 2
redistribution, 10, 15, 80
reduction, 77
refining, 1
repeatability, 39
research, 3, 79
rheumatoid arthritis, 3
room temperature, 37

S

salts, 27, 37
sample, 37, 39, 66
scaling, 48, 80, 81
series, 2, 11, 28, 37, 40, 45, 65
serine, 4
software, 84
solid phase, 3, 16, 29, 33, 43
solid state, 47, 64, 65, 68, 73, 76
solid-state, viii, 3, 8, 27, 31, 37, 40, 45, 46, 65, 73, 77, 81
solvent, 16, 27, 33, 35, 64, 80, 81
solvent molecules, 16, 27, 34, 35, 80, 81
species, 9, 16, 28, 63
spectra, 3, 8, 37, 38, 39, 78, 81
spectral analysis, 77
spectroscopy, 39
spectrum, 11, 37, 39, 40, 42, 43, 44, 45, 46, 69, 70, 72, 73, 74, 75, 77
stability, 9, 11

stabilization, 33, 52, 64
standard deviation, 48, 80
strain, 11
stretching, 37, 40, 45, 47
subtraction, 39
suspensions, 3
symmetry, 32, 75
synthesis, vii, 2, 81
systematic, vii, 3
systems, 1, 10, 65

T

targets, 1, 2
theoretical, 8, 15, 29, 31, 32, 33, 34, 37, 40, 47, 48, 52, 63, 64, 65, 77, 79, 80
theory, viii, 9, 12, 14, 15, 29, 32, 34, 48, 52, 54, 57, 63, 64, 79, 80
threonine, 4
time, 3, 37
toxin, vii, 2
transfer, 10
transition, 39, 40, 47, 69, 70, 72, 73, 75, 76, 77, 78
tripeptide, 65
Trp, vii, 2
tryptophan, 4, 29
tyrosine, 4, 34

U

Unrestricted Hartree-Fock (UHF), 9, 12, 14, 15, 16, 17, 18, 19, 20, 21, 22, 23, 24, 25, 26, 27, 30, 32, 34, 48, 58, 59, 79, 80
urinary, 2

V

validity, 34, 37
valine, 4, 64
values, 15, 28, 29, 31, 34, 39, 47, 48, 64, 65, 80
van der Waals, 2
vector, 39
vibration, 40
vibrational, viii, 3, 8, 37, 47, 76, 79, 80, 81
viruses, vii, 2

W

water, 2, 40

X

X-ray, vii, 3, 8, 16, 29, 34, 37, 40, 47, 48, 52, 65, 77, 79, 80
X-ray diffraction, vii, 8, 16, 37, 40, 48, 52, 65, 79, 80
X-ray diffraction data, 16, 52, 80